ENVIRONMENTAL RIGHTS FOR FUTURE GENERATIONS

This book presents readers with a new theory and practice of international human rights law that is designed to improve its protection of the environmental rights of future generations. Arguing that international law is currently unable to safeguard future generations from foreseeable environmental harm, Bridget Lewis proposes that the law needs to be reformed in the interests of achieving intergenerational justice. The book draws on different theories of intergenerational responsibility to articulate a fresh approach, revising both substantive principles of environmental rights and procedural rules of admissibility and standing. It looks at several case studies to explore how the proposed new approach would apply in relation to contemporary environmental challenges like fracking, deep seabed mining, nuclear energy, decarbonisation and geoengineering.

Bridget Lewis is a human rights scholar whose research explores various issues at the intersection of the environment and human rights. Lewis's first book, *Environmental Human Rights and Climate Change: Current Status and Future Prospects* (2018), critically examined the right to a healthy environment in the context of the climate crisis. Her recent research explores the human rights of future generations and the emerging trend of children's climate litigation.

Environmental Rights for Future Generations

BRIDGET LEWIS
Queensland University of Technology

CAMBRIDGE
UNIVERSITY PRESS

CAMBRIDGE
UNIVERSITY PRESS

Shaftesbury Road, Cambridge CB2 8EA, United Kingdom

One Liberty Plaza, 20th Floor, New York, NY 10006, USA

477 Williamstown Road, Port Melbourne, VIC 3207, Australia

314–321, 3rd Floor, Plot 3, Splendor Forum, Jasola District Centre, New Delhi – 110025, India

103 Penang Road, #05–06/07, Visioncrest Commercial, Singapore 238467

Cambridge University Press is part of Cambridge University Press & Assessment, a department of the University of Cambridge.

We share the University's mission to contribute to society through the pursuit of education, learning and research at the highest international levels of excellence.

www.cambridge.org
Information on this title: www.cambridge.org/9781009574198

DOI: 10.1017/9781009574174

When citing this work, please include a reference to the DOI 10.1017/9781009574174

First published 2025

Cover image: Darling basin – stock photo/Southern Lightscapes-Australia/Getty Images

A catalogue record for this publication is available from the British Library

Library of Congress Cataloging-in-Publication Data
NAMES: Lewis, Bridget author
TITLE: Environmental rights for future generations / Bridget Lewis, Queensland University of Technology.
DESCRIPTION: Cambridge, United Kingdom ; New York, NY : Cambridge University Press, 2025. | Includes bibliographical references and index.
IDENTIFIERS: LCCN 2025018108 (print) | LCCN 2025018109 (ebook) | ISBN 9781009574198 hardback | ISBN 9781009574211 paperback | ISBN 9781009574174 epub
SUBJECTS: LCSH: Environmental policy–International cooperation | Environmental law | Human rights–International cooperation | Intergenerational relations
CLASSIFICATION: LCC GE170 .L495 2025 (print) | LCC GE170 (ebook) | DDC 363.7–dc23/eng/20250710
LC record available at https://lccn.loc.gov/2025018108
LC ebook record available at https://lccn.loc.gov/2025018109

ISBN 978-1-009-57419-8 Hardback

For Hugh, Marigold, Angus and Felicity

Contents

Table of Cases

INTERNATIONAL COURT OF JUSTICE

Advisory Opinion on the Legal Consequences of the Construction of a Wall in the Occupied Palestinian Territories (International Court of Justice, 9 July 2004)

Barcelona Traction, Light and Power Company, Limited (Belgium v Spain) [1970] ICJ Rep 3

Case Concerning Pulp Mills on the River Uruguay (Argentina v Uruguay) [2010] ICJ Rep 14

East Timor (Portugal v Australia) [1995] ICJ Rep 90

Gabčíkovo-Nagymaros Project (Hungary/Slovakia) (Separate Opinion of Judge Weeramantry) [1997] ICJ Rep 88

Legal Consequences of the Construction of a Wall in the Occupied Palestinian Territory (Advisory Opinion) [2004] ICJ Rep 136

Legality of the Threat or Use of Nuclear Weapons (Advisory Opinion) [1996] ICJ Rep 226

Legality of the Threat or Use of Nuclear Weapons (Advisory Opinion) (Dissenting Opinion of Judge Weeramantry) [1996] ICJ Rep 429

Legality of the Use by a State of Nuclear Weapons in Armed Conflict (Advisory Opinion) [1996] ICJ Rep 66

Nuclear Tests Case (Australia v France) (Interim Protection Order) [1973] ICJ Rep 99

Nuclear Tests Case (Australia v France) (Judgment) [1974] ICJ Rep 253

Nuclear Tests Case (New Zealand v France) (Interim Protection Order) [1973] ICJ Rep 135

Nuclear Tests Case (New Zealand v France) (Judgment) [1974] ICJ Rep 457

Obligations of States in Respect of Climate Change (Request for Advisory Opinion) (International Court of Justice, 29 March 2023)

Request for an Examination of the Situation in accordance with Paragraph 63 of the Court's Judgment of 20 December 1974 in the Nuclear Tests Case (New Zealand v France) (Order) [1995] ICJ Rep 288

EUROPEAN COURT OF HUMAN RIGHTS

Andreou v Turkey [2010] Eur Court HR, Application No 45653/99
Brincat and Others v Malta [2014] Eur Court HR, Application Nos 60908/11, 62110/11, 62129/11, 62312/11, 62338/11
Budayeva and Others v Russia [2008] Eur Court HR, Application Nos 15349/02, 21166/02, 20058/02, 11673/02 and 15343/02
Cordella and Others v Italy [2019] Eur Court HR, Application Nos 54414/13 and 54264/15
Di Sarno et al v Italy [2012] Eur Court HR, Application No 30765/08
Duarte Agostinho and Others v Portugal and 32 Others [2024] Eur Court HR, Application No 39371/20
Dubetska v Ukraine [2011] Eur Court HR, Application No 30499/03
Fadeyeva v Russia [2005] Eur Court HR, Application No 55723/00
Fagerskiold v Sweden [2008] Eur Court HR, Application No 37664/04
Giacomelli v Italy [2006] Eur Court HR, Application No 59909/00
Greenpeace Nordic and Others v Norway (Petition) (Eur Court HR, 15 June 2021)
Handyside v the United Kingdom [1976] Eur Court HR, Application No 5493/72
Hatton and Others v the United Kingdom [2003] Eur Court HR, Application No 36022/97
Keenan v the United Kingdom [2001] Eur Court HR, Application No 27229/95
Kolyadenko and Others v Russia [2012] Eur Court HR, Applications Nos 17423/05, 20534/05, 20678/05, 23263/05, 24283/05 and 35673/05
Kovačić et al v Slovenia [2008] Eur Court HR, Applications Nos 44574/98, 45133/98 and 48316/99
Kyrtatos v Greece [2003] Eur Court HR, Application No 41666/98
LCB v United Kingdom [1999] Eur Court HR, Application No 23413/94
Lopez Ostra v Spain [1994] Eur Court HR, Application No 16798/90
Mansur Yalçın and Others v Turkey [2014] Eur Court HR, Application No 21163/11
Öneryildiz v Turkey [2004] Eur Court HR, Application No 27785/10
Osman v the United Kingdom [1998] Eur Court HR, Application No 23452/94

Soering v the United Kingdom [1989] Eur Court HR, Application No 14038/88
Soubeste and Others v Austria and 11 Other States [2022] Eur Court HR, Application No 31925/22
Taskin v Turkey [2004] Eur Court HR, Application No 46117/99
Verein KlimaSeniorinnen Schweiz v Switzerland [2024] Eur Court HR, Application No 53600/20

INTER-AMERICAN COURT AND COMMISSION

Case of Human Rights Defender et al v Guatemala (Inter-American Court of Human Rights, 28 August 2014)
Case of the Pueblo Bello Massacre v Colombia (Inter-American Court of Human Rights, 31 January 2006)
Comunidad Yanomami (Inter-American Commission on Human Rights, Case No 7615, 5 March 1985)
Emérita Montoya González v Costa Rica (Admissibility) (2006) Inter-American Commission on Human Rights, Report No 48/96
Environment and Human Rights: State Obligations in Relation to the Environment in the Context of the Protection and Guarantee of the Rights to Life and to Personal Integrity (Advisory Opinion) [2017] Inter-American Court of Human Rights OC-23/17
González et al ("Cotton Field") v Mexico (Preliminary Objection, Merits, Reparations, and Costs) [2009] Ser C 205
Jessica Lenahan (Gonzales) v United States [2011] Inter-Am Court HR, Case No 12.626
Request for an Advisory Opinion on the Climate Emergency and Human Rights Submitted to the Inter-American Court of Human Rights by the Republic of Colombia and the Republic of Chile (Inter-American Court of Human Rights, 9 January 2023)
Valle Jaramillo et al v Colombia (Merits, Reparations, and Costs) [2008] Inter-Am Court HR, Case No 12.415, Ser C No 192
Yakye Axa Indigenous Community v Paraguay (Merits, Reparations and Costs) [2005] Inter-Am Court HR, Ser C No 125

AFRICAN COURT AND COMMISSION

Article 19 v Eritrea (2007) African Commission on Human and Peoples' Rights, Communication No 275/2003

Law Society of Zimbabwe and Others v Zimbabwe [2016] African Commission on Human and Peoples' Rights, Communication No 321/2006

Social and Economic Rights Action Center and the Center for Economic and Social Rights v Nigeria [2002] African Commission on Human and Peoples' Rights, Case No 155/96

HUMAN RIGHTS COMMITTEE

Aalbersberg v Netherlands [2006] Human Rights Committee, Communication No 1440/2005

Andersen v Denmark [2010] Human Rights Committee, Communication No 1868/2009

Beydon v France [2005] Human Rights Committee, Communication No 1400/2005

Brun v France [2006] Human Rights Committee, Communication No 1453/2006

Daniel Billy and Others v Australia [2022] Human Rights Committee, Communication No 3624/2019

EW et al v The Netherlands [1993] Human Rights Committee, Communication No 429/1990

Ioane Teitiota v New Zealand [2019] Human Rights Committee, Communication No 2728/2016

Ivan Kitok v Sweden [1988] Human Rights Committee, Communication No 197/1985

Ominayak (Lubicon Lake Band) v Canada [1990] Human Rights Committee, Communication No 167/1984

OTHER INTERNATIONAL CASES

Advisory Opinion on Request submitted to the Tribunal by the Commission of Small Island States on Climate Change and International Law [2024] International Tribunal for the Law of the Sea, Case No 31

Award between the United States and the United Kingdom Relating to the Rights of Jurisdiction of United States in the Bering's Sea and the Preservation of Fur Seals (1893) XXVIII Reports of International Arbitral Awards 263

Sacchi et al v Argentina [2021] Committee on the Rights of the Child, Communication No 104/2019

SC and GP v Italy [2019] Committee on Economic, Social and Cultural Rights, Communication No 22/2017
Trail Smelter Case (USA v Canada) (1941) III RIAA 1905

NATIONAL CASES

Ali v Pakistan [2016] Supreme Court of Pakistan Constitutional Petition No ___ / I of 2016
Demanda Generaciones Futuras v Minambiente [2018] (Supreme Court of Colombia, Case No 11001-22-03-000-2018-00319-01, 5 April 2018)
ENvironnement JEUnesse v Canada [2021] QCCA Quebec Court of Appeal 500-09-028523-199, 1871
ENvironnement JEUnesse v Procureur General du Canada [2019] Superior Court of Québec 500-06-000955-183
ENvironnement JEUnesse v Procureur General du Canada (Application) (Superior Court of Québec, 26 November 2018)
Held et al v State of Montana (Notice of Appeal) [2023] Supreme Court of Montana DA 23-0575
Held et al v State of Montana (Opinion) [2024] Supreme Court of Montana (Case No DA 23-0575) 2024 MT 312
Held et al v State of Montana (Order) [2023] Montana First Judicial District Court (Lewis and Clark County) Cause No CDV-2020-307
Juliana v US (Complaint) (US District Court (District of Oregon, Eugene Division), Case No 6:15-cv-01517-TC
Juliana v US [2016] US District Court (District of Oregon, Eugene Division) Case No 6:15-cv-01517-TC
Juliana v US [2020] United States Court of Appeals for the Ninth Circuit 18-36082
Kochale and Others v Lake Turkana Wind Power Ltd and Others [2021] Environment and Land Court at Meru, Kenya, Case No 163/2104
Minister for the Environment v Sharma (2022) [2022] FCAFC 35 (15 March 2022)
Minors Oposa v Secretary of the Department of Environment and Natural Resources (1994) 33 ILM 173
Neubauer et al v Germany [2021] Federal Constitutional Court 1, Case Nos BvR 2656/18, 1 BvR 78/20, 1 BvR 96/20, 1 BvR 288/20
Pabai Pabai and Anor v Commonwealth of Australia (Federal Court of Australia, VID622/2021)
Pandey v Union of India [2019] National Green Tribunal, Principal Bench, New Delhi 187/2017

Sharma by Her litigation Representative Sister Marie Brigid Arthur v Minister for the Environment (2021) [2021] FCA 560 (Federal Court of Australia, 27 May 2021)

State of the Netherlands v Urgenda Foundation [2018] Hague Court of Appeal, Case No ECLI:NL:RBDHA:2015:7145

State of the Netherlands v Urgenda Foundation [2019] Supreme Court of The Netherlands, Case No 19/00135 (ECLI:NL:RBDHA:2015:7145)

Statnett SF et al v Sør-Fosen sijte et al [2021] Supreme Court of Norway, Case No HR-2021-1975-S

Union Hidalgo, ECCHR, ProDESC v Electricité de France SA [2024] Paris Court of Appeal

US v Juliana [2024] US Court of Appeals 9th Cir No 24-684

Waratah Coal Pty Ltd v Youth Verdict Ltd and Others (No. 6) [2022] Land Court of Queensland 21

Woodpecker et al v South Korea (Constitutional Court of South Korea, 2022Hunma864)

Table of International Conventions

African Charter on Human and Peoples' Rights 1981
American Convention on Human Rights 1969
Charter of the United Nations 1945
Convention Concerning the Protection of the World Cultural and Natural Heritage 1972
Convention on Access to Information, Public Participation in Decision-Making and Access to Justice in Environmental Matters 1998
Convention on Biological Diversity 1992
Convention on the Elimination of All Forms of Discrimination against Women 1979
Convention on the Prevention and Punishment of the Crime of Genocide 1948
Convention on the Rights of the Child 1989
First Optional Protocol to the International Covenant on Civil and Political Rights 1966
International Convention on the Elimination of All Forms of Racial Discrimination 1965
International Covenant on Civil and Political Rights 1966
International Covenant on Economic, Social and Cultural Rights 1966
Joint Convention on the Safety of Spent Fuel Management and on the Safety of Radioactive Waste Management 2014
Optional Protocol to the Convention on the Rights of the Child on a Communications Procedure 2011
Optional Protocol to the International Covenant on Economic, Social and Cultural Rights 2008
Paris Agreement 2015

Abbreviations

BECCS	bio-energy with carbon capture and storage
BTEX	benzene, toluene, ethylbenzene and xylene
CERD	Convention on the Elimination of All Forms of Racial Discrimination
CRC	Convention on the Rights of the Child
ECHR	European Convention on Human Rights
EEZ	Exclusive Economic Zone
EPA	Environmental Protection Agency
FPIC	free, prior and informed consent
IAEA	International Atomic Energy Agency
ICCPR	International Covenant on Civil and Political Rights
ICESCR	International Covenant on Economic, Social and Cultural Rights
ICJ	International Court of Justice
IHRL	international human rights law
IPCC	Intergovernmental Panel on Climate Change
ISA	International Seabed Authority
NDC	nationally determined contribution
PPT	Permanent People's Tribunal
SAI	sulphur aerosol injection
SRM	solar radiation management
UN	United Nations
UNCLOS	United Nations Convention on the Law of the Sea
UNDRIP	United Nations Declaration on the Rights of Indigenous Peoples
UNESCO	United Nations Educational, Scientific and Cultural Organization
UNFCCC	United Nations Framework Convention on Climate Change
VOC	volatile organic compounds

1

Introduction

1.1 SETTING THE SCENE

The world is facing unprecedented environmental challenges. We have exceeded a number of critical planetary boundaries and are facing crises of climate change, biodiversity loss and pollution. Without an urgent change of course, we risk potentially catastrophic shifts in the systems that sustain human and non-human life on our planet. We have arrived at this point because of a historic attitude of entitlement towards Earth's natural resources and, more recently, a failure of political and economic processes to respond to a crescendo of warnings from the scientific community.

Collectively, our past and ongoing actions have set a trap for future generations. At best, they will bear the responsibility of repairing the environmental damage already wrought at the expense of pursuing other opportunities for their own flourishing. At worst, they face the prospect of a planet which cannot supply them with adequate and safe food, water and shelter and which is subjected to regular catastrophes of extreme weather, in a context of increased inequity, disease and conflict.

These profound environmental and social challenges have been a focus of human rights discourse for some time now, and the potential for environmental damage and destruction to affect human rights is plain to see. Climate change is the biggest threat to human rights the modern world has ever confronted, a multiplier of existing inequalities and vulnerabilities, and a potential existential danger if not tackled urgently. As our understanding of the natural world improves, guided by the knowledge of Indigenous peoples and enhanced by scientific endeavours, we have come to appreciate that all harm to the environment affects human rights.

Just as the climate crisis has presented an unprecedented problem for science, economics and politics, so too is it a massive challenge for human rights law. Twenty years ago, the first seeds of a climate movement based in human rights were planted with an Inuit petition to the Inter-American Commission on Human Rights. Enthusiasm for human rights-based approaches to climate change has gathered momentum, and we have seen climate litigation using human rights law in many jurisdictions. This trend has coincided with a growing recognition of environmental human rights, most recently by the UN General Assembly in its resolution on the right to a clean, healthy and sustainable environment.

However, it has taken some time for international law to catch up with the instinctive understanding that human rights and the environment are intricately linked. There have been cases where environmental conditions have been found to represent a breach of human rights, and some expert bodies have sought to articulate the relationships between human rights and the environment, but the protection of environmental rights at the international level is piecemeal at best. And, despite this progress, these legal frameworks have struggled to address environmental harm that is predicted to impact human rights in the future.

At the time of its inception, human rights law offered such immense potential to prevent the sorts of indignities and violence that the world had witnessed during the Second World War. Since then, we have seen countless examples of governments failing to live up to that optimistic vision of a world based on respect for inherent human dignity; yet through those times there has been a sense that human rights law at least had the language to express our collective admonition, if not also the tools to punish the wrongdoers and prevent future violations. Climate change has tested our faith in human rights, as we have seen governments and private companies knowingly act in ways that undermine human security, agency and dignity.

In trying to understand why international human rights law (IHRL) has come up short in dealing with long-term climate change, I have been drawn to the work of Dryzek and Pickering and their concept of ecological reflexivity.[1] The laws, institutions and processes of human rights have not been able to adapt to the environmental crisis that is unfolding, even though we can clearly see the way that environmental conditions and the ecological state of the planet impact on people's human rights. Human rights law needs more 'ecological reflexivity' – a capacity to adapt to ensure its ongoing relevance and usefulness as planetary conditions change.

[1] John S Dryzek and Jonathan Pickering, *The Politics of the Anthropocene* (Oxford University Press, 2018).

This book presents an account of the main reasons why IHRL has not adequately addressed the environmental threats facing future generations. These reasons go back to the original conceptualisation of IHRL as a framework for protecting citizens against the harmful excesses of their own governments. The early human rights treaties focused on restraining governmental power, with the addition of some positive expectations to advance the enjoyment of all human rights. The legal enforcement of these obligations remained limited. As the realities of environmental and climate impacts have emerged, IHRL has continued to adopt individualised, backwards-looking models of liability, instead of more forward-focused, collective and cooperative responsibilities. It remains constrained by outdated geographic and temporal limitations on state responsibilities.

Because state duties have been defined relative to their spheres of jurisdiction and control, human rights litigation was for a long time limited to cases brought by people against their own governments. There were some exceptions to this, but these were usually in circumstances where a person found themselves within the territory or under the control of a foreign state. There were occasional attempts to litigate a transboundary human rights impact, such as a shooting across an international border or foreign military activity, but these were only successful where it was shown that the state had exercised control resulting in a direct impact on the applicant's human rights.

More recently, human rights case law and commentary have started to open up to the idea of extraterritorial obligations. Now, if a state's actions have a foreseeable impact on the rights of people in other countries, it is more likely that their human rights obligations will be engaged. However, despite this expansion of human rights duties into other geographic spaces, they remain limited in terms of time. For a state to be responsible for protecting against a future threat to human rights, it is usually necessary to show that the risk is specifically identifiable, directly linked to the state's actions and set to manifest imminently. It is still very difficult to establish that a state has a legal obligation to protect against long-term impacts on human rights. At the international level at least, litigation is not possible on behalf of future generations, even where there is strong evidence that their rights will be negatively affected by a state's actions.

This book is an attempt to chart a different course that might overcome this problem. It seeks to pinpoint the reasons why IHRL has difficulty addressing future environmental harm and then proposes a number of recommendations to help solve these problems. The barriers identified relate to the nature of states' obligations, to whom they are owed and what they require, as well as the issue of who can bring legal action to enforce IHRL. Further questions of how an allegation might be proved and what causation looks like in the context of future human rights impacts are also addressed.

The analysis in the book concentrates on human rights litigation, rather than other forms of human rights advocacy, supervision or education. There are several reasons for this focus. To begin with, the limitations of IHRL have been born out of the interpretations of human rights courts and tribunals. Human rights cases are the setting where we have tested the limits of the law and are therefore a key space where new paths can be opened up. Amending international human rights treaties is a much more difficult undertaking but is arguably unnecessary if changes can evolve through judicial interpretation. Perhaps more importantly, the effectiveness of human rights laws is often judged through our ability to enforce them. If rights cannot be enforced against a government whose actions threaten their enjoyment, we might question the value of the laws that claim to protect those rights.

The discussion in this book also focuses on international law, rather than domestic or regional human rights law, although some lessons are taken from domestic and regional litigation. There are a few reasons for this. For one, IHRL sets the tone for national laws by imposing obligations on countries and establishing standards that governments are expected to meet. Adapting IHRL to include more future-focused duties should therefore serve as a driver of change at the local level. IHRL is also positioned as a last resort for enforcement when local mechanisms fail. Complaints to the human rights treaty bodies are available after the exhaustion of local remedies. IHRL is therefore the last chance to address a problem that can't be fixed elsewhere, so plugging the gaps in international law can be a way of safeguarding rights in the absence of domestic law. And finally, many of the issues that threaten the environmental rights of future generations are governed by international rules and regulations, such as the UN Framework Convention on Climate Change or the UN Convention on the Law of the Sea's regime for deep seabed mining, or influenced by principles of international environmental law. The book seeks to reach a deeper understanding of how these different sets of norms might interact and influence the rights of future generations.

With this in mind, the book attempts to understand and overcome the reasons why litigating future generations' human rights is so difficult in IHRL. By unpacking the specific aspects of human rights litigation that are sticking points for future generations' cases, we can then target these areas in seeking new approaches. The result is a number of suggested changes to the way we think about the building blocks of litigation, including jurisdiction, standing, admissibility, causation and proof. The book also suggests a revised approach to defining states' obligations, which is inspired by intergenerational equity and informed by the precautionary principle and due diligence. The result is a set of duties and principles which, it is hoped, could make IHRL more future-focused and operationalise the general notion that states owe obligations to protect the rights of future generations. Before providing a short overview of the argument presented in the book, a few preliminary matters need to be clarified.

1.2 DEFINITION OF FUTURE GENERATIONS

One of the challenges of securing the rights of future generations has been the fact that for a long time we haven't had an agreed definition of who the term applies to. The means and effectiveness of protecting their rights will vary considerably depending on whether the class includes only unborn people or also extends to children. This can also influence whether we think it is necessary to protect future generations' rights at all.

John Knox, former UN Special Rapporteur on the environment and human rights, has argued that we can address the impacts of climate change and other environmental harms on future generations if we work to secure the rights of all people who will be alive at a certain point in the future, say the year 2100. He argues we can do this by focusing on the rights of children today.[2] Others have argued that we can achieve intergenerational justice (including for generations not yet born) by working to improve children's enjoyment of human rights.[3]

While it's clear that children overlap with future generations, protecting future generations' rights is distinct from protecting children's rights for many reasons. As will be explained in Chapter 4, a key reason for distinguishing future generations from children is that children alive today do not necessarily have the same interests as people who will be born in the future. In relation to enforcement through litigation, relying on children to act as representatives of future generations is also problematic. Aoife Nolan has argued that amalgamating children's rights with future generations' rights could distort both sets of rights and the duties that go with them, undermining our efforts to advance rights for both groups.[4] Aoife Daly has also argued

[2] John H Knox, 'Constructing the Human Right to a Healthy Environment' (2020) 16(1) *Annual Review of Law and Social Science* 79.

[3] Karin Arts, 'Children's Rights and Climate Change' in Claire Fenton-Glynn (ed), *Children's Rights and Sustainable Development* (Cambridge University Press, 1st ed, 2019) 216.

[4] Aoife Nolan, 'Children and Future Generations' Rights before the Courts: The Vexed Question of Definitions' (2024) 13(3) *Transnational Environmental Law* (in press); see also Aoife Nolan, 'Children's Rights and Climate Change at the UN Committee on the Rights of the Child: Pragmatism and Principle in Sacchi v Argentina', *EJIL: Talk!* (20 October 2021)

that children and future generations are distinct classes and that there is no overlap between the two and that future generations are those who are not yet born.[5] The *Maastricht Principles on the Human Rights of Future Generations* similarly defines future generations as 'those generations that do not yet exist but will exist and who will inherit the earth. Future generations include persons, groups and peoples'.[6]

Part of the difficulty in defining future generations is the fluidity of generations. Each generation overlaps with those which come before and after it, and there will be many differences in the needs, interests, capabilities and vulnerabilities of different groups between and within each generation. Consequently, we should be careful not to consider future generations as one single, abstract group. The Maastricht Principles address this in the Preamble, which says: 'Human generations exist within an unbroken continuum that is continually renewed and redefined as untold new members join the living human community. Any treatment of human generations and their respective rights must recognize and reflect this continuum'.[7]

A similar approach was recently endorsed by the UN Committee on the Rights of the Child, in its General Comment on children's rights and the environment. In addressing the relationship between children's rights and the environment, the Committee recognised:

> the principle of intergenerational equity and the interests of future generations . . . While the rights of children who are present on Earth require immediate urgent attention, the children constantly arriving are also entitled to the realization of their human rights to the maximum extent. Beyond their immediate obligations under the Convention with regard to the environment, States bear the responsibility for foreseeable environment-related threats arising as a result of their acts or omissions now, the full implications of which may not manifest for years or even decades.[8]

In keeping with this idea of a continuum of constantly arriving and renewing generations, this book does not place any temporal restrictions on the class of future generations. Instead, it defines them as persons who do not yet exist, but

<www.ejiltalk.org/childrens-rights-and-climate-change-at-the-un-committee-on-the-rights-of-the-child-pragmatism-and-principle-in-sacchi-v-argentina>.

5 Aoife Daly, 'Climate Competence: Youth Climate Activism and Its Impact on International Human Rights Law' (2022) 22(2) *Human Rights Law Review* 1.

6 'Maastricht Principles on the Human Rights of Future Generations' 1 <www.ohchr.org/sites/default/files/documents/new-york/events/hr75-future-generations/Maastricht-Principles-on-The-Human-Rights-of-Future-Generations.pdf>.

7 Ibid Preamble, para III.

8 Committee on the Rights of the Child, 'General Comment No. 26 (2023) on Children's Rights and the Environment, with a Special Focus on Climate Change' [11].

will exist in the future. There is no limit on how far into the future the definition extends, but it is noted that, as future people become more remote in time, the challenges of knowing and protecting their interests become more difficult. The definition encompasses both individual members of future generations and groups who will exist in the future, but does not view future generations as a collective themselves. Thus, it seeks to find ways to protect both individual and group rights of future generations.

It is important to note that 'future generations' is not intended to include embryos or foetuses. The choice to define future generations as persons who 'do not yet exist but will exist in the future' rather than 'unborn people' is a deliberate one to minimise any potential application to foetuses or distortion of future generations' rights into rights of unborn children. The recognition of the rights of people who will exist in the future does not imply a right for any currently unborn person to be born, and does not in any way override the right of women or people capable of becoming pregnant to make their own sexual and reproductive choices.

The Maastricht Principles make a similar point, explaining that:

> Nothing in these Principles recognises any rights of human embryos or foetuses to be born nor does it recognise an obligation on any individual to give birth to another. These Principles may not be construed as accepting any interferences with the bodily autonomy of women girls, and others who can become pregnant, including their actions and decisions around pregnancy or abortion and other sexual and reproductive health and rights.[9]

Thus, this book looks for ways to meaningfully protect the human rights of people who will exist in the future, while accepting that future people are constantly arriving in the present, and without distinguishing between different groups of future people.

1.3 RATIONALE FOR PROTECTING FUTURE GENERATIONS' ENVIRONMENTAL RIGHTS

The argument which runs through this book begins from the premise that IHRL should protect the environmental rights of future generations. It presents several reasons for this, beginning with the fact that there are many environmental problems that have serious long-term effects on human rights. As the case studies in this book will illustrate, many of the greatest environmental problems facing the world are rightly understood as human rights

[9] Ibid 4(c).

problems as well. Because the environmental damage is long-term, so too is the human rights impact.

For example, climate change threatens a wide range of human rights. This has been acknowledged for over a decade now, with the UN High Commissioner for Human Rights recognising the human rights risks of climate change as early as 2009.[10] A large body of work has identified the impacts of climate change on human rights and the role of human rights law in addressing this global problem.[11] Rising temperatures due to climate change will increase future generations' exposure to extreme weather events like bushfires and floods, impact on their food and water security, and put them at risk of new and exacerbated health problems. It is clear, therefore, that future generations will be forced to deal with the long-term human rights consequences of present-day environmental decision-making.

What is more, the risks and burdens facing future generations become increasingly serious the longer we delay taking action, due to the ways in which climate change 'feeds itself' through exponentially worsening feedback loops and the potential of reaching critical planetary tipping points.[12] The fact that future generations will suffer interferences with their human rights as a result of climate change is recognised as a significant intergenerational injustice, which is exacerbated by the fact that future generations have had no role in causing the problem and have no voice in deciding on laws and policies to address it.[13]

[10] Office of the High Commissioner for Human Rights, *Report of the Office of the United Nations High Commissioner for Human Rights on the Relationship between Climate Change and Human Rights* (15 January 2009).

[11] Meinhard Doelle, 'Climate Change and Human Rights: The Role of the International Human Rights in Motivating States to Take Climate Change Seriously' (2004) 1(2) *Macquarie Journal of International and Comparative Environmental Law* 179; John Knox, 'Climate Change and Human Rights Law' (2009) 50 *Virginia Journal of International Law* 163; John H Knox, 'Linking Human Rights and Climate Change at the United Nations' (2009) 33 *Harvard Environmental Law Review* 477; Ole Pedersen, 'Climate Change and Human Rights: Amicable or Arrested Development?' (2010) 1(2) *Journal of Human Rights and the Environment* 236; Sumudu Atapattu, *Human Rights Approaches to Climate Change* (Routledge, 2016); Bridget Lewis, *Environmental Human Rights and Climate Change: Current Status and Future Prospects* (Springer, 2018); Ottavio Quirico and Mouloud Boumghar (eds), *Climate Change and Human Rights* (Routledge, 2015).

[12] For a helpful discussion of the justice implications of these feedback loops, see Henry Shue, *The Pivotal Generation: Why We Have a Moral Responsibility to Slow Climate Change Right Now* (Princeton University Press, 2021).

[13] Bridget Lewis, 'The Rights of Future Generations within the Post-Paris Climate Regime' (2018) 7(1) *Transnational Environmental Law* 69; Bridget Lewis, 'Human Rights Duties towards Future Generations and the Potential for Achieving Climate Justice' (2016) 34(3) *Netherlands Quarterly of Human Rights* 206.

Our responses to climate change also raise issues with respect to future generations' environmental rights. For instance, decisions about how to transition to renewable energy involve many choices about where to build wind and solar farms, the outcome of which can have long-term environmental impacts if not properly regulated. 'Green' technologies like solar panels and electric vehicles require critical minerals which are in increasing demand, triggering a new mining boom even as fossil fuel extraction decreases. A new frontier for such exploration is the deep seabed, where deposits of rare earth elements could be exploited to capitalise on this growing market. The process of extracting minerals from the seabed could threaten the marine life which lives there – ecosystems about which relatively little is known – and disturb sediments with unknown consequences for water quality and sea life more broadly. The desire to cut energy-based emissions also raises the prospect of greater reliance on nuclear energy as an alternative to fossil fuels, with associated concerns about the location of facilities and the management of waste products.

Geoengineering, the deliberate manipulation of the Earth's systems, has been on the table for some time as a possible strategy for counteracting anthropogenic climate change.[14] The possibility of 'fixing' climate change through technology has obvious appeal. Furthermore, most of the scientific modelling of pathways that could succeed in keeping global warming within the agreed limit of 2°C assumes that some form of 'negative emissions' technology like carbon capture and storage is going to be employed.[15] It is therefore very likely that decisions will be made in the near future to pursue or scale up various geoengineering strategies.

While geoengineering promises to help us avoid the dangers of climate change, it also presents its own risks and uncertainties. Geoengineering might have significant negative impacts on atmospheric, geologic and marine systems, with possibly severe consequences for future generations. Once we have committed to a particular geoengineering strategy, the consequences of failure or sudden deactivation of that technology could be extremely serious.[16]

[14] Royal Society (Great Britain), *Geoengineering the Climate: Science, Governance and Uncertainty* (The Royal Society, 2009).

[15] Katherine Calvin et al, *Climate Change 2023: Synthesis Report. Contribution of Working Groups I, II and III to the Sixth Assessment Report of the Intergovernmental Panel on Climate Change (Summary for Policymakers)* (Intergovernmental Panel on Climate Change (IPCC), 25 July 2023) 22–23.

[16] Andy Parker and Peter J Irvine, 'The Risk of Termination Shock from Solar Geoengineering' (2018) 6(3) *Earth's Future* 456.

There are other risks associated with the technology options we choose. Decisions about which technologies we research, field-test and scale up, the locations of those activities and the ways in which they will be monitored and regulated create the potential for path dependence and lock-in, which may foreclose other options for future generations.[17] Once locked in, the challenge to maintain the chosen technology and prevent potential catastrophe from 'termination shock' will fall to future generations. The application of human rights law to decisions regarding geoengineering therefore deserves greater consideration.

Alongside the climate crises, the planet is on the brink of an extinction disaster, with biodiversity loss occurring at an increasing rate. There are many separate but compounding causes for this, including habitat loss and fragmentation, over-exploitation through hunting and fishing, pollution and climate change. Because of the interconnectedness of ecosystems, the decline or loss of one species can have drastic ramifications for other related species.

Biodiversity loss must also be understood as a human rights issue. The protection of biodiversity has instrumental value for the enjoyment of a wide range of rights. We need only consider the importance of insect pollinators for the survival of agricultural industries to recognise the key role which these species play in our economies and in ensuring food security. The emerging theory of One Health also explains the critical role which biodiversity plays for human health. It has been defined as:

> An integrated, unifying approach that aims to sustainably balance and optimize the health of people, animals and ecosystems. It recognizes the health of humans, domestic and wild animals, plants, and the wider environment (including ecosystems) are closely linked and inter-dependent.[18]

Beyond this instrumental value that biodiversity has for particular human rights, the inherent value of the non-human world is increasingly understood to have significance within human rights. Many jurisdictions now recognise more ecocentric versions of the right to a healthy environment, through which the well-being of the environment itself must be protected, not just its role in advancing human well-being. For Indigenous peoples, certain species and ecosystems can have important cultural and spiritual value, such that a loss of biodiversity has direct impacts on the well-being of both individuals and

[17] Catriona McKinnon, 'Sleepwalking into Lock-in? Avoiding Wrongs to Future People in the Governance of Solar Radiation Management Research' [2018] *Environmental Politics* 1.

[18] 'Joint Tripartite (FAO, OIE, WHO) and UNEP Statement: Tripartite and UNEP Support OHHLEP's Definition of "One Health"'.

communities. The widespread threats to biodiversity can therefore be understood as a significant threat to human rights as well.

Once we accept that these environmental problems are also human rights problems, responding to them adequately requires that we recognise and address their human rights consequences. And because these environmental problems are long-term issues, our human rights-based responses must also address the human rights consequences for future generations.

This book also argues that IHRL must address future generations' environmental rights because it is just and right to do so. The intrinsic link between human rights and justice is at the foundation of modern human rights law, reflected in fundamental concepts of universality, equality and non-discrimination. If this commitment to equality and justice is to mean anything in the context of environmental rights, human rights protections should extend to future generations, given the very serious ways in which their rights will be impacted by future environmental harms. It is also appropriate that IHRL reflects the broad perspectives of the people and communities that it seeks to protect. This entails a need to be open to non-Western and Indigenous perspectives regarding intergenerational responsibility. Failing to do so would conflict with the ideas of equality and justice that human rights law claims to uphold. For these various reasons, the book argues that IHRL should protect the environmental rights of future generations.

1.4 THE SHORTCOMINGS OF EXISTING LAW

There are several reasons, however, why IHRL fails to meet this expectation. These flow from both the way that rights and duties are defined, as well as the rules and procedures through which they are enforced. One of the chief reasons for international human rights law's inability to address the rights of future generations is its use of imminence as a defining factor of states' obligations. Imminence is used in three different ways, though these are not entirely distinct and often bleed into one another.

First, imminence has a role in determining standing to bring a complaint under most international and regional human rights laws. In most systems, a person can bring a claim to a human rights body if they have been affected by an act or omission that has an actual or imminent impact on their rights. If the impact on their rights is more remote, they will not normally be considered a victim and will lack standing to bring a complaint.

The second sense in which imminence has been applied is in relation to positive duties of states to protect human rights. This is most relevant in relation to rights like the right to life, which imposes both negative and positive duties on states. States have a negative duty not to kill or cause people's deaths, but also a positive duty to address life-threatening conditions by taking appropriate protective steps. However, this positive duty has been interpreted to apply only when threats to life are imminent.[19] This is intended to place some limits on what states are expected to do, but it has had the effect that some threats to life, though they may be foreseeable or even likely, are deemed not to require a response from the state.[20]

A third role that imminence has played in human rights law is in relation to causation. To establish that a state has violated its legal obligations, it is usually necessary to show that the negative human rights impact was caused by the state's act or omission. This is usually simple enough where harm has already occurred, but becomes more challenging where the harm is anticipated to occur in the future. Again, imminence has been used as the standard here. Courts can be satisfied that a causal link exists between an act and its imminent consequences. However, cases alleging more distant harm have been unsuccessful because courts have concluded that there is time for an intervening act to occur which might sever the causal chain.[21]

To further complicate matters, the meaning of imminence is not consistent, nor is the weight given to it across these three areas. It is often conflated with ideas of immediacy or directness, which focus on the specificity of an impact rather than its timeframe. A human rights breach that is expected to affect an identified individual will usually be easier to litigate than one whose victims cannot yet be known. The foreseeability of a future impact is also critical, but again this can be interpreted differently in terms of whether knowledge must be objective or subjective, specific or general. In some cases, general knowledge of foreseeable danger can be enough to trigger a state's duties, without a need to know the specifics of when or how that danger will materialise.[22]

Despite these variations in the way imminence has been defined and applied, it has remained a consistent presence in international and regional human rights litigation. Any claim that seeks to address human rights impacts

[19] *Osman v the United Kingdom* [1998] Eur Court HR 23452/94.

[20] For a very useful analysis of imminence and positive duties, see Justine Bell-James and Briana Collins, 'Human Rights and Climate Change Litigation: Should Temporal Imminence Form Part of Positive Rights Obligations?' (2022) 13(1) *Journal of Human Rights and the Environment* 212.

[21] *Ioane Teitiota v New Zealand* [2019] Human Rights Committee Communication No 2728/2016.

[22] See e.g. *Öneryildiz v Turkey* [2004] Eur Court HR [GC] 27785/10.

for future generations will be confronted with challenges in terms of establishing standing, relevant state duties and causal connections.

IHRL is also limited by the way it has approached environmental rights, which is typically through an anthropocentric rather than ecocentric framing. The development of environmental rights in IHRL has mostly been through a process of 'greening' existing rights, rather than recognising a standalone right to an environment of any particular quality. Thus, we have a body of case law which has found violations of rights to life, to private and family life, to an adequate standard of living or to health based on environmental harm.[23] In recent years, the United Nations General Assembly has recognised the right to a clean, healthy and sustainable environment.[24] However, the resolution does not recognise a standalone right to an environment of a particular quality, but instead articulates the essential interdependence of human rights and the environment. In doing so, the resolution does not create any new obligations for states. Indeed, as a General Assembly resolution it can only ever have persuasive force at best and cannot create legally binding duties.

This anthropocentric framing of the right to a healthy environment, through which the environment is protected because of its instrumental value in supporting other human rights and not for its own sake, limits its potential use for future generations. As noted earlier, human rights cases generally take the form of claims about specific, direct harms to identifiable individuals. They are more difficult to make out when the harm is more remote or diffuse. A right which protected the environment for its own sake would arguably be easier to apply in a future-oriented way, since a breach could be based on foreseeable environmental harm, even if the 'human' consequences of that harm were more difficult to predict.

1.5 A NEW APPROACH

This book proposes a new approach which aims to tackle both the substantive and procedural challenges in upholding environmental rights for future

[23] See John H Knox, 'Greening Human Rights', *openDemocracy* (14 July 2015) <www.opendemocracy.net/en/openglobalrights-openpage/greening-human-rights>; Bridget Lewis, 'Environmental Rights or a Right to the Environment?: Exploring the Nexus between Human Rights and Environmental Protection' (2012) 8 *Macquarie Journal of Comparative and International Environmental Law* 36; Karrie Wolfe, 'Greening the International Human Rights Sphere? Environmental Rights and the Draft Declaration of Principles on Human Rights and the Environment' (2003) 9(1) *Appeal: Review of Current Law and Law Reform* 45.

[24] United Nations General Assembly, 'Resolution 76/300: The Human Right to a Clean, Healthy and Sustainable Environment'.

generations. It offers a new theory of intergenerational responsibility for environmental rights which is tailored to IHRL and oriented towards protecting future generations' rights. This is informed by various theories of intergenerational equity and justice, as well as the worldviews of Indigenous peoples and non-Western thinking about rights and duties. In particular, it integrates the work of Edith Brown Weiss by joining intergenerational equity with the tripartite duties to respect, protect and fulfil human rights within international law. This new model also incorporates the precautionary principle and concept of due diligence to push for a more protective, proactive interpretation of states' obligations. This, it is argued, helps to give content to states' duties to prevent harm to future generations where the risks are foreseeable.

To enable the legal enforcement of these more future-oriented duties, the proposed reforms start by recognising that future generations are holders of human rights, both individually and collectively, and that they possess the full range of rights that any person living today enjoys. The conditions upon which state duties will be engaged are inspired by recent work around extraterritorial duties: where a state exercises control over an activity which is likely to impact on future generations' rights, the state will owe duties with respect to those rights, provided that the impact was foreseeable to the state at the time it exercised that control. The model also includes provision for standing via representative claims, so that future generations' rights can be legally protected.

The final section of the book then presents three case studies which road-test this approach. First, it looks at two forms of unconventional resource extraction – fracking and deep seabed mining. The second case study looks in detail at climate change and our responses to it, considering the potential future impacts of decarbonisation, geoengineering and adaptation activities, as well as the harms of global warming itself. The third case study considers the implications of nuclear energy for future generations' environmental rights and investigates how the proposed new approach might apply to nuclear cases.

Each case study is an example of a current activity that creates risks for future generations, but each also presents slightly different issues. This helps to illustrate the strengths and potential weaknesses of the proposed new model and shows how legal arguments might need to be framed in litigation on behalf of future generations. Together, they help us to envision what might be achieved through a reformed version of IHRL based on intergenerational responsibility.

There are endless ways in which our actions today can and will impact on the ability of future generations to enjoy their human rights. Even with the best available science and the best of intentions, it's inevitable that we will still make decisions – either knowingly or unwittingly – that have negative consequences for them. However, we can limit the dangers of this by at least ensuring that our human rights laws do not allow those negative consequences purely because they will occur in the future. This book seeks to acknowledge that we owe duties to protect the rights of future generations, and proposes a way of legally enforcing those duties. By doing so, it hopes to bring future generations' environmental rights into focus and to elevate them in state decision-making so that they can no longer be ignored.

2

The Rationale

Why International Human Rights Law Should Recognise the Environmental Rights of Future Generations

2.1 INTRODUCTION

The body of international human rights law has a long history and is rooted in even older philosophical arguments and theories about who has human rights and why. As the next chapters will illustrate, this body of law has so far not addressed future generations' human rights, and has only recently expanded to give some protection to environmental rights. Some might very well ask why international human rights law (IHRL) should recognise and protect the environmental rights of future generations. After all, IHRL already covers a wide range of rights and arguably struggles to keep up with all the ways these rights are abused. Perhaps it would be best to leave future generations' interests to dedicated institutions and mechanisms? Given the complexity of striking a balance between current and future interests, might not these questions best be left for states to determine for themselves? Maybe we should focus our efforts within IHRL on improving implementation and enforcement of rights and duties that are already accepted? Indeed, there is work to be done in many spaces, but this chapter seeks to lay out a rationale for the central argument of the book: that IHRL should protect the environmental rights of future generations.

The moral case for protecting future generations' interests is well made by various theories of intergenerational equity and intergenerational justice, which emphasise the fact that future people are equally entitled to have their basic needs fulfilled and their interests protected. A sense of intergenerational responsibility is also part of the worldviews of many Indigenous peoples, derived from kin-like relationships with the natural world and strong connections with both ancestors and descendants. Two fundamental principles of IHRL itself are equality and justice. It therefore ought to respect and reflect these broader understandings of responsibility, ensure that future generations benefit from equal protection, and contribute to intergenerational justice.

In recent times, the international community has taken some steps towards the recognition of both environmental rights and intergenerational responsibility. The General Assembly and Human Rights Council have recognised the right to a clean, healthy and sustainable environment.[1] These resolutions acknowledge the interconnectedness of the environment and all human rights, and the importance of protecting biodiversity and ecosystems. By incorporating the notion of sustainability, they also echo the principles of international environmental law, which seeks to ensure natural resources are managed in a way which supports future generations' needs.

In 2023, the Maastricht Principles on the Human Rights of Future Generations were released and endorsed by international experts, including several current UN Special Rapporteurs on human rights.[2] The Principles explain that recognising future generations' rights is an 'essential dimension of humankind's duty to uphold the inherent dignity, equality and inalienable rights of all'.[3] They articulate the obligations of governments and the international community as a whole to respect, protect and fulfil future generations' human rights through appropriate mechanisms. The framers of the Maastricht Principles have encouraged human rights treaty bodies to use the Principles when interpreting the rights contained in their respective international conventions. This chapter expands on this call by setting out the various reasons why IHRL should adapt to provide better protections for future generations' rights, specifically environmental rights.

Drawing on different schools of moral and political philosophy, the chapter outlines the many arguments why current generations are morally obliged to address long-term environmental problems that threaten the interests and well-being of future generations. These arguments can be found in theories

1 United Nations General Assembly, 'Resolution 76/300: The Human Right to a Clean, Healthy and Sustainable Environment'; United Nations Human Rights Council, 'Resolution 48/13: The Human Right to a Clean, Healthy and Sustainable Environment' <https://documents-dds-ny.un.org/doc/UNDOC/GEN/G21/289/50/PDF/G2128950.pdf?OpenElement>.

2 'Maastricht Principles on the Human Rights of Future Generations' 1 <www.ohchr.org/sites/default/files/documents/new-york/events/hr75-future-generations/Maastricht-Principles-on-The-Human-Rights-of-Future-Generations.pdf>.

3 Ibid Preamble.

based on distributive justice, equity, planetary trusts and commons, and capabilities. Duties to future generations are also found within many Indigenous knowledge systems. For example, the Seventh Generation Principle of Native American peoples and the philosophies of Aboriginal and Torres Strait Islander peoples in Australia illustrate the custodial sense of responsibility that many Indigenous peoples feel towards the natural world and future generations.

After outlining the various explanations for a moral obligation to future generations, the chapter explains in Section 2.3 why IHRL should be a location for this obligation to be performed. It considers the relationship between human rights and justice articulated in the founding documents of IHRL, which recognise that the full enjoyment of human rights is the foundation of peace and justice in the world. Given this fundamental link between human rights and justice, it is reasonable to expect that environmental rights (and human rights law more generally) will advance justice, including intergenerational justice.

Human rights law is also fundamentally based on the concepts of equality and universality, according to which all people are equal in dignity and equally entitled to the full enjoyment of human rights. Human rights must therefore be enjoyed without discrimination, and I argue here that this must extend to prevent discrimination between generations. Drawing together these fundamental concepts of justice and equality, it is argued that IHRL must include the rights of future generations.

This is also a natural conclusion of the increased recognition of the right to a clean, healthy and sustainable environment. As explained in more detail in Chapter 3, the right to a healthy environment has been widely recognised in domestic laws and constitutions, and has recently been recognised by UN bodies. This trend has solidified recognition of the relationship between human rights and the environment and increasingly brings the non-human world into the gaze of human rights law. The emphasis on a 'sustainable' environment within these developments makes it necessary to think about environmental rights beyond the current generations – the enjoyment of a healthy environment by future generations must be protected if 'sustainability' is to have any meaning as an objective of human rights law. Drawing these propositions together, the chapter argues that future generations' rights are implied within environmental rights protections.

As the chapter argues, recognising future generations' rights is consistent with a more inclusive, democratised and decolonised understanding of human rights law, which integrates a plurality of perspectives on rights and responsibilities. As will be explored in more detail in Chapter 5, a reimagining of environmental rights which is influenced by Indigenous and non-Western epistemologies can accommodate the rights of people of all times and places.

This chapter articulates the rationale behind this book overall and establishes the overarching objective for the analysis in later chapters, which is to explore the reasons why future generations' environmental rights are not adequately protected under existing laws and to identify theoretical and pragmatic ways this can be remedied. To begin with, though, let us turn to consider the various theoretical accounts for a moral responsibility towards future generations.

2.2 THEORIES OF INTERGENERATIONAL RESPONSIBILITY

The introductory chapter of this book highlighted some of the very serious long-term consequences that can flow from environmentally damaging activities. It made the connection between this kind of environmental harm and human rights, arguing that environmental rights need to have a future orientation in order to address these harms effectively. While it might be factually evident that environmental problems will affect future people, that fact on its own does not establish a reason for us to do anything about it, especially since many of those environmentally damaging activities are likely to have benefits for us in the short term. To make the moral case for addressing future generations' rights, we need to turn to theories of intergenerational responsibility.

There are many moral philosophers who have tackled the question of what we owe to future people: how should we adjust our actions today, if at all, to account for the interests of future generations? Many prominent scholars have argued that we do owe moral duties towards future people. These theories take various shapes and forms, and this section will outline the main variations. It is noted, however, that there is considerable overlap and cross-fertilisation of ideas among different theories, and they are not all easily amenable to categorisation. Terminology in this space can vary as well. Most theories tend to use the language of either intergenerational justice or intergenerational equity. As a generic term to cover the range of approaches, intergenerational responsibility is adopted here as the preferred language, as it is less obviously linked to any particular philosophical traditions and is capable of encompassing a range of approaches.

In addition to Western traditions of intergenerational justice or equity, intergenerational responsibility also forms a core part of many Indigenous worldviews. The knowledge and belief systems of many First Nations peoples integrate obligations of care and custodianship towards the natural world with a sense of responsibility towards their descendants and ancestors.

The following discussion gives an overview of the various ways in which we can conceive of a moral responsibility towards future generations.

2.2.1 *Justice to Future Generations*

At the heart of many approaches to intergenerational responsibility is a fundamental desire to treat future generations fairly. Particularly in more recent times, a body of work has emerged from the recognition that our past and present practices have created burdens for future people which are unjust.

Several specific theories of intergenerational responsibility can be traced back to John Rawls' theory of distributive justice, which aims to share society's benefits and burdens fairly.[4] Rawls argued that society could agree on a just distribution if we applied a 'veil of ignorance' to our own individual position – that is, if we ignored each individual's own advantages and characteristics, including the point in time at which each is born.[5] Once in this 'original position', where each person could come from any era of time, society could agree to a principle of 'just savings' to determine how to use our resources based on what is fair to all generations.[6] As Rawls put it: '[e]ach generation must not only preserve the gains of culture and civilization, and maintain intact those just institutions that have been established, but it must also put aside in each period of time a suitable amount of real capital accumulation'.[7] Drumbl explains how this concept of just savings applies to environmental concerns: '[I]f people in the "original position" did not know what generation they would be born into, then persons would take better care of the environment if there were a risk that they could be born into a generation whose predecessors had desecrated the environment or emptied it of its resources'.[8]

Rawls' theory has been critiqued, including in relation to the intergenerational application of the just savings principle. His theory of justice envisages the distributive rules that society would agree to if all parties were placed in the hypothetical 'original position' under the 'veil of ignorance', free from the

4 John Rawls, A *Theory of Justice* (Harvard University Press, 1971); John Rawls, *Justice as Fairness: A Restatement*, ed Erin I Kelly (Harvard University Press, 2001); Malgosia Fitzmaurice, 'Intergenerational Equity Revisited' in Isabelle Buffard et al (eds), *International Law between Universalism and Fragmentation* (Brill Nijhoff, 2008) 195, 199.

5 Rawls, A *Theory of Justice* (n 4) 136–142.

6 Ibid 285.

7 Ibid.

8 Mark A Drumbl, 'Poverty, Wealth, and Obligation in International Environmental Law' (2001) 76(4) *Tulane Law Review* 843, 919; Fitzmaurice (n 4) 199.

biases that come with the advantages and disadvantages of their actual lives. It is therefore a form of social contract theory, dependent on rational choices made to benefit all in a way which all consider to be fair. Critics like Dierksmeier and Dauenhauer have pointed out the difficulty of extending this contractarian approach to persons who do not physically exist and who therefore cannot participate in the negotiation and whose interests can more easily be ignored.[9] In his later work, Rawls argued that: 'the correct principle, then, is one the members of any generation (and so all generations) would adopt as the principle they would want preceding generations to have followed'.[10]

But the criticisms of Rawls' theory have remained, with scholars like Tremmel arguing that a theory of intergenerational justice must have a moral foundation, and not one based purely on rational contractarianism and 'sheer human self-interest'.[11] Others have pointed to the fact that social contract theories implicitly rely on reciprocity, but this is not possible in relation to future generations.[12] A strict distributive justice approach based on the idea of the social contract clearly struggles to explain the moral imperative to act fairly towards future generations.

Dryzek and Pickering argue that a new approach to justice is needed for the Anthropocene, since 'injustice looms large in a disrupted planet'.[13] They articulate a new theory of planetary justice that includes global, intergenerational and non-human dimensions. This approach needs to operate at the level of the Earth's systems, considering what kind of Earth system we are leaving for future generations, and recognising that responsibilities sit not just with individuals but also with collectives.[14]

Henry Shue has articulated the reasons why traditional theories of intergenerational justice can be of limited use for the problem of climate change and argued that the question of how much current generations should be

9 Claus Dierksmeier, 'John Rawls on the Rights of Future Generations' in Jörg Tremmel (ed), *Handbook of Intergenerational Justice* (Edward Elgar, 2006) 72; Bernard Dauenhauer, 'Response to Rawls' in Richard A Cohen and James L Marsh (eds), *Ricoeur as Another: The Ethics of Subjectivity* (State University of New York Press, 2002) 203; Fitzmaurice (n 4) 201–203.

10 Rawls, *Justice as Fairness* (n 4) 160.

11 Jörg Tremmel (ed), *Handbook of Intergenerational Justice* (Edward Elgar, 2006) 10; Fitzmaurice (n 4) 203.

12 Fitzmaurice (n 4) 203–204; Edward Page, *Climate Change, Justice and Future Generations* (Edward Elgar, 2006).

13 John S Dryzek and Jonathan Pickering, *The Politics of the Anthropocene* (Oxford University Press, 2018) 58.

14 Ibid 70.

doing remains contested. The compounding nature of climate change means that sequential generations are not facing comparable challenges, but future generations' burdens get exponentially worse the longer current generations delay acting. Furthermore, the narrow approach of considering what one generation must do for the next is not enough when the timescale of climate change is so massive. An approach which merely looks to conserve resources, for example, is unlikely to be adequate when future generations will face much greater costs and have much different needs as a result of our actions.

These critiques of Rawls' approach have helped to refine theories of intergenerational responsibility, particularly those which are oriented towards the specific problems of the Anthropocene. And, in spite of the critiques, Rawls' work has remained influential, as can be seen in the work of Edith Brown Weiss. Brown Weiss's theory of intergenerational equity seeks to unpack the reasons why we owe obligations to future generations and what those obligations entail.

2.2.2 *Edith Brown Weiss and Intergenerational Equity*

Intergenerational equity has become a commonly invoked principle of international environmental law, and is therefore an important theory to consider in our search for a moral rationale for protecting the environmental rights of future generations. A particularly influential theory of intergenerational equity is that developed by Edith Brown Weiss. She argued that current generations have an obligation to leave the planet in no worse condition than they acquired it. To do this, present generations must act in accordance with three guiding principles: conservation of options, conservation of quality and conservation of access.[15] Brown Weiss's theory draws on several other concepts related to intergenerational responsibility to establish the fundamental premise that present generations are obliged to respect the rights of future generations and deal with their interests equitably.

As a starting point, Brown Weiss identified that 'we, the human species, hold the natural environment of our planet in common with all members of our

[15] Edith Brown Weiss, *In Fairness to Future Generations: International Law, Common Patrimony and Intergenerational Equity* (Transnational, 1989); Edith Brown Weiss, 'Our Rights and Obligations to Future Generations for the Environment' (1990) 84 *American Journal of International Law* 198; Edith Brown Weiss, 'Climate Change, Intergenerational Equity, and International Law' (1985) 9 *Vermont Journal of Environmental Law* 615; Edith Brown Weiss, 'In Fairness to Future Generations and Sustainable Development' (1992) 8(19–26) *American University International Law Review*.

species: past generations, the present generation, and future generations'.[16] Because all human beings are connected through their ongoing relationship with the Earth, it follows that all generations have an equal place with respect to the natural system, and the interests of future generations are no less important than those of present generations. According to Brown Weiss, this common and equal interest gives rise to a partnership between past, present and future generations in which all humans are linked through the natural world.[17]

From this starting point, Brown Weiss then looked to determine principles which could guide present generations in how exactly to implement their intergenerational responsibilities. She used four criteria to help determine appropriate guiding principles:

1. Principles should not authorise present generations to exploit natural resources to the exclusion of future generations, nor impose unreasonable burdens on present generations to meet indeterminate future needs;
2. No principle should require us to predict future values, but should be sufficiently flexible to allow future generations to achieve their goals according to their values;
3. Principles should be clear in their application to foreseeable situations;
4. Principles should be shared by different cultural traditions and be generally acceptable across different economic and political systems.[18]

In following this approach to articulate appropriate principles, Brown Weiss drew on Rawls' theory of distributional justice, which Fitzmaurice calls the 'philosophical background' of Brown Weiss's doctrine.[19] Using Rawls' concept of the 'veil of ignorance', she argued that 'future generations would want to inherit the Earth in as good a condition as did their ancestors and with at least comparable access to its resources'.[20] This general idea that we must leave the Earth in no worse condition than we found it is the cornerstone of Brown Weiss's theory of intergenerational equity.

Drawing these various concepts together, Brown Weiss articulated her three normative principles of conservation of options, quality and access. Conservation of options guarantees to future generations a degree of diversity and flexibility comparable to that enjoyed by previous generations. This

16 Brown Weiss, 'Our Rights and Obligations to Future Generations for the Environment' (n 15) 198–199; Brown Weiss, *In Fairness to Future Generations* (n 15).

17 Brown Weiss, 'In Fairness to Future Generations and Sustainable Development' (n 15) 21.

18 Ibid 23.

19 Fitzmaurice (n 4) 218.

20 Brown Weiss, 'In Fairness to Future Generations and Sustainable Development' (n 15) 21.

includes diversity of both natural and cultural resources, so that future generations are not unduly restricted in their options.[21] Conservation of quality requires each generation to pass on the environment in no worse condition than they received it. Brown Weiss explained that options and quality need to be distinguished as two separate principles, since future generations will not only need diversity of natural resources to allow them options, but also adequate quality of those resources to support their needs and avoid undue burdens of environmental degradation.[22] The third principle, conservation of access, ensures each generation has equitable ability to access the 'legacy of past generations' and is obliged to safeguard the same access for future generations.[23] These principles, Brown Weiss emphasises, do not compel present generations to take any particular action or manage natural resources in any particular way, but rather guide them to choose policies and actions which will conserve options, quality and access for future generations.[24]

While not directing any particular action, Brown Weiss did identify several activities that are likely to impinge on future generations' rights and which require consideration. These include the generation of waste that cannot be restricted either spatially or temporally (including nuclear and other hazardous waste); loss of biodiversity; ozone depletion; contamination or depletion of freshwater supplies; and destruction of cultural heritage, libraries and archives.[25] All these activities engage present generations' intergenerational obligations and should be addressed in a way which ensures conservation of options, quality and access. She suggests some strategies for doing this, including through representation in decision-making (e.g. via an envoy or ombudsman); policies aimed at maintaining systems, archives and investments that will benefit future generations; and long-term scientific research and development to improve natural resource management and address long-term environmental threats.[26]

2.2.3 *Planetary Trusts and the Commons*

A key feature of Brown Weiss's theory is the concept of a planetary trust and the associated notion that present generations hold the Earth as trustees or

[21] Ibid 22.
[22] Ibid.
[23] Ibid 23; Fitzmaurice (n 4) 207.
[24] Brown Weiss, 'In Fairness to Future Generations and Sustainable Development' (n 15) 23.
[25] Ibid 24, 26.
[26] Ibid 25–26.

stewards. She notes that this idea is common across many human traditions.[27] Each generation holds the Earth's natural resources 'as a trust passed to us by our ancestors for our benefit, but also to be passed on to our descendants for their use'.[28] In this respect, future generations can be thought of as both beneficiaries of the trust and trustees for the generations to come.

There are other theories that advocate for intergenerational responsibilities based on the idea that natural resources or environmental areas are enjoyed by all species (or at least all humans) in common and passed on from one generation to the next. This 'common patrimony' gives rise to obligations with respect to its management and exploitation for the collective good. These obligations are commonly framed in terms of a trust or guardianship arrangement, though there are differences in terminology. The 'common heritage of humankind' usually includes both environmental and cultural resources. The 'global commons' is typically restricted to certain environmental areas, such as ocean areas beyond state jurisdictions and the atmosphere (though views differ as to what areas are appropriate to include within global commons). Theories of an 'intergenerational trust' are more specifically oriented towards responsibilities across time, rather than just between members of a collective with mutual interests.[29]

A version of the concept of an environmental trust was applied in the *Bering Fur Seal Arbitration* in 1893, where the tribunal found that natural resources are held in public trust for humankind, which necessitates regulatory measures for the use of those resources beyond national jurisdictions.[30] The decision thereby draws the link between common patrimony and public trusts. In some legal systems, a public trust is recognised which places the government in the role of trustee, obliged to safeguard certain beneficial public interests. These might include access to fishing or navigation in marine areas, ecological protection or scenic beauty. Legal enforcement options are available where the government fails to perform its duties under the public trust.

Fitzmaurice points out that the concept of a public trust can be problematic, especially when we attempt to apply it for the benefit of future generations, as the protected uses may conflict and the trust does not necessarily

27 Ibid 20.

28 Ibid.

29 Catherine Redgwell, *Intergenerational Trusts and Environmental Protection* (Juris Publishing; Manchester University Press, 1999).

30 *Award between the United States and the United Kingdom relating to the Rights of Jurisdiction of United States in the Bering's Sea and the Preservation of Fur Seals* (1893) XXVIII Reports of International Arbitral Awards 263; Fitzmaurice (n 4) 212.

establish a hierarchy which can help to balance the different beneficial uses.[31] Without a clear principle of intergenerational equity to guide agencies and legislators, future generations' interests may take a back seat to those of present generations. It is for this reason that Brown Weiss uses the planetary trust as a starting point for establishing intergenerational responsibilities, which she then unpacks through the three principles of intergenerational equity.

2.2.4 *Capabilities and Rights for Future Generations*

Several theories of intergenerational responsibility take a capabilities-[32] or interests-based approach,[33] arguing that we should modify our actions today out of concern for the interests of future generations. For example, Simon Caney explains that we must address 'harmful short-termism' which threatens the fundamental rights and interests of future people.[34] Van Dijk uses capabilities theory to identify the set of capabilities that future generations will need, and argues that we should ensure that our actions do not undermine their ability to enjoy those capabilities.

A related approach is to recognise that future generations will have basic rights which trigger obligations for current generations. Generally speaking, rights-based approaches to intergenerational responsibility proceed on the basis that future generations will be entitled to the same catalogue of rights that we currently enjoy, and consequently we have obligations in relation to those rights. In that sense, rights-based approaches also draw on the concept of equality, since there is no reason to treat future generations any differently when recognising their rights. The precise content of the obligations that go with these rights is less settled, but at the very least it would likely include an obligation not to act knowingly in a way which would prevent future people from being able to enjoy their rights. It may also extend as far as creating conditions which would facilitate the full enjoyment of rights in the future.

31 Fitzmaurice (n 4) 215–216.

32 Nicky van Dijk, 'The Capability Approach as a Road Map for Re-thinking Intergenerational Justice' in Jan Linehan and Peter Lawrence (eds), *Giving Future Generations a Voice: Normative Frameworks, Institutions and Practice* (Edward Elgar, 2021) 42.

33 Joel Feinberg, 'The Rights of Animals and Future Generations' (1971).

34 Simon Caney, 'Cosmopolitan Justice, Rights and Global Climate Change' (2006) 19(2) *Canadian Journal of Law and Jurisprudence* 255; Simon Caney, 'Climate Change, Human Rights and Moral Thresholds' in Stephen Humphreys (ed), *Human Rights and Climate Change* (Cambridge University Press, 2009) 69; Michael K MacKenzie, *Institutional Design and Sources of Short-Termism* (Oxford University Press, 2016); Iñigo González-Ricoy and Axel Gosseries, 'Designing Institutions for Future Generations' in Iñigo González-Ricoy and Axel Gosseries (eds), *Institutions for Future Generations* (Oxford University Press, 2016) 3.

The chief criticism against the concept of future generations' rights is that future people do not exist and therefore cannot possess rights (or anything else). Beckerman and Pasek rejected the notion that future generations could be rights holders, arguing that, even if we hold that current generations owe obligations, this does not give rise to rights. Their view of rights is non-symmetrical: rights can give rise to obligations but not the other way around.[35] Macklin argued that once a person becomes an 'actual person' they are entitled to human rights, but they cannot possess them beforehand.[36] Gosseries explained this in terms of what he called the 'right-bearer contemporaneity' requirement: rights and duties can only exist between contemporaries.[37]

Another well-known criticism is what Parfit termed the 'non-identity' problem. Parfit argued that our current actions determine not only whether people will be born in the future but also who those people will be, since we will change the conditions in which people come to exist. Consequently, we cannot be said to owe any obligations to future generations, since any change to our actions today would cause those people not to exist in the future.[38] D'Amato has endorsed this argument, explaining that by acting to protect the future environment we ensure that 'different people will be born from those who would have been born if we had not intervened in the environment'.[39] This then creates 'Parfit's paradox': 'How can we owe a duty to future persons if the very act of discharging that duty wipes out the very individuals to whom we allegedly owed that duty? Our attempted environmental altruism will prevent the birth of the precise beneficiaries of our altruism'.[40] According to D'Amato, the sense of obligation to preserve the environment for the benefit of future people is 'nonsensical' because the act of preservation will destroy those same people.[41] And if there can be no obligations owed to future people, it follows that it is equally incoherent to talk of their having rights.

35 Wilfred Beckerman and Joanna Pasek, *Justice, Posterity and the Environment* (Oxford University Press, 2001).

36 Ruth Macklin, 'Can Future Generations Correctly Be Said to Have Rights?' in E Partridge (ed), *Responsibilities to Future Generations: Environmental Ethics* (Prometheus, 1981) 151, 152.

37 Axel Gosseries, 'On Future Generations' Future Rights' (2008) 16(4) *Journal of Political Philosophy* 446, 456.

38 Derek Parfit, *Reasons and Persons* (Clarendon, 1984); see also Anthony D'Amato, 'Do We Owe a Duty to Future Generations to Preserve the Global Environment?' (1990) 84(1) *American Journal of International Law* 190; Fitzmaurice (n 4) 208.

39 D'Amato (n 38) 191.

40 Ibid.

41 Ibid 193.

Edith Brown Weiss countered these criticisms by characterising future generations' rights as group rights. For Brown Weiss, it did not matter that the identities of future individuals are unknown, or that we do not know how many such individuals will exist, or that their identities might change, because the rights are possessed collectively by the generation as a whole.[42] For Brown Weiss, there was no question that future generations possess rights. Such rights flow from the entitlements owed to them under the trust relationship and each generation's part in the intertemporal entity of human society.[43] She also pointed to the fact that framing their entitlements as rights 'focuses discussion on the welfare of generations, what each generation is able to have and enjoy, in a way that obligations [alone] cannot'.[44] She further argued that these rights and obligations should be recognised and protected through legal frameworks, and advocated for enforcement through a guardian, ombudsman or similar representative.[45] By framing future generations' rights as group rights, Brown Weiss overcomes the non-identity problem, while also connecting to the sense that humans share the natural world in partnership with each other and other species.[46]

These criticisms against the idea of future generations' rights can also be addressed if we adopt an interest theory of rights. Such theories posit that, because we can conceive that future generations will have interests, it is coherent to conclude that they will have rights. The interest theory of rights was developed by scholars including Joseph Raz, who explained that human rights are those things in which we have a significant interest. As Raz explained it, a person has a right to *x* when *x* is a fundamental interest that is weighty enough to impose obligations on others.[47] Caney gives a similar explanation, saying that 'we ascribe rights to protect highly valued interests (such as liberty of conscience, association, and expression) and our standard ascription of rights is guided by our account of what persons' most important interests are'.[48]

42 Brown Weiss, 'Our Rights and Obligations to Future Generations for the Environment' (n 15) 204–206; Brown Weiss, 'In Fairness to Future Generations and Sustainable Development' (n 15) 24–25.

43 Brown Weiss, 'Our Rights and Obligations to Future Generations for the Environment' (n 15) 202; Brown Weiss, 'In Fairness to Future Generations and Sustainable Development' (n 15) 20.

44 Brown Weiss, 'Our Rights and Obligations to Future Generations for the Environment' (n 15) 204.

45 Ibid 202–203; Brown Weiss, 'In Fairness to Future Generations and Sustainable Development' (n 15) 24.

46 Fitzmaurice (n 4) 207.

47 Joseph Raz, *The Morality of Freedom* (Clarendon, 1986) 166.

48 Caney, 'Cosmopolitan Justice, Rights and Global Climate Change' (n 34) 255, 259; Simon Caney, 'Human Rights, Climate Change and Discounting' (2008) 17(4) *Environmental Politics* 536.

Feinberg has argued that anything can have rights where it is capable of having interests, that is, where it can have a good or well-being of its own which should be protected by legal or moral rules.[49] From this basis he has argued that animals and future generations can possess rights, but not 'mere things', which have no capacity to have desires or aims. Van Dyke's account of human rights also aligns with interest theory as he asserts that rights are things which are essential to the pursuit of our most basic interests and needs.[50]

The interest theory can be used to overcome Parfit's non-identity problem. A number of scholars have argued that the fact we do not know the identities of future people, and even the fact that those identities may change, does not change the fact that future people (whoever they are) will have human rights.[51] As Feinberg says: '[t]he identity of the owners of these interests is now necessarily obscure, but their interest-ownership is crystal clear, and that is all that is necessary to certify the coherence of present talk about their rights'.[52] Bell has similarly argued that 'the indeterminacy of the identity of future persons does not offer any grounds for denying that they will have human rights'.[53] What is more, our inability to know the identities of future people does not change the reality that our actions today will determine the quality of life that they enjoy. Neither does the non-identity problem address the morally intuitive sense that there is something wrong with knowingly acting in a way which will result in a poorer quality of life for people in the future. The interest theory of rights therefore provides an explanation of both the basis for and content of our responsibilities towards future generations.

While these approaches offer different understandings of why we owe responsibilities to future generations, and especially what those responsibilities look like in practice, what they have in common is a sense of moral duty to consider their interests and to act in a way which helps to promote and protect those interests. The majority of Western scholarship, especially when considering the complex problems presented by the Anthropocene and climate

[49] Feinberg (n 33) 5–6.

[50] Vernon Van Dyke, 'Collective Entities and Moral Rights: Problems in Liberal-Democratic Thought' (1982) 44(1) *The Journal of Politics* 21, 23.

[51] Joel Feinberg, 'The Rights of Animals and Unborn Generations' in E Partridge (ed), *Responsibilities to Future Generations: Environmental Ethics* (Prometheus, 1981) 139; Derek Bell, 'Does Anthropogenic Climate Change Violate Human Rights?' (2011) 14(2) *Critical Review of International Social and Political Philosophy* 99; Caney, 'Cosmopolitan Justice, Rights and Global Climate Change' (n 34); Robert Elliot, 'The Rights of Future People' (1989) 6 *Journal of Applied Philosophy* 159.

[52] Feinberg (n 51).

[53] Bell (n 51) 107.

change in particular, agrees that there is a moral obligation for current generations to care for the rights of future generations.

2.2.5 *Indigenous Worldviews*

The proposition that we should care for the interests of future generations and act accordingly is not a new concept, nor is it limited to Western philosophical traditions. Versions of this thinking can be found in the philosophies and knowledge systems of Indigenous peoples around the world. For many peoples, a sense of obligation to other generations (past and future) and a collective duty to care for the natural world are fundamental to their spiritual, cultural and social practices and beliefs.

For instance, the Seventh Generation principle is part of the worldview of some First Nations of North America. The principle requires that any decision or action should safeguard the health of the land for seven generations to come.[54] According to this principle, when a Chief makes a decision for their people, they must consider how that decision would impact on their descendants.[55]

The Seventh Generation principle is commonly associated with the Haudenosaunee (Iroquois) people, and it formed part of the Great Law of the Haudenosaunee Confederacy, the oral constitution of the Five (later Six) Nations of the Haudenosaunee.[56] Other nations have similar custodial worldviews, flowing from their belief that the Creator gave the natural world to them to use, to benefit from and to protect. Clarkson et al explain the origins of the principle in these terms:

> It is said that we are placed on the earth (our Mother) to be the caretakers of all that is here. We are instructed to deal with the plants, animals, minerals, human beings and all life, as if they were a part of ourselves. Because we are a part of Creation, we cannot differentiate or separate ourselves from the rest of the earth. The way in which we interact with the earth, how we utilize the plants, animals and the mineral gifts, should be carried out with the seventh generation in mind. We cannot simply think of ourselves and our survival; each generation has a responsibility to 'ensure the survival for the seventh generation'.[57]

54 Bob Joseph, 'What Is the Seventh Generation Principle?' <www.ictinc.ca/blog/seventh-generation-principle>; Haudenosaunee Confederacy, 'Values', *Haudenosaunee Confederacy* <www.haudenosauneeconfederacy.com/values>.

55 Haudenosaunee Confederacy (n 54).

56 Joseph (n 54).

57 Linda Clarkson, Vern Morrissette and Gabriel Ragallet, *Our Responsibility to the Seventh Generation* (International Institute for Sustainable Development, 3 October 2001) 12 <www.iisd.org/publications/our-responsibility-seventh-generation>.

The Seventh Generation principle is not only applicable to environmental matters, however, but is also critical to ensuring the longevity of First Nations' culture and language.[58] It came to influence all facets of Indigenous societies and customs. As Clarkson et al explain,

> This basic law that was the driving force behind the development of Indigenous culture became reflected in the institutions and systems of Indigenous people … the extended family systems, the clan system, decision-making through consensus, division of labour respecting the respective roles of the clans and based upon need, survival and family structure all contribute to sharing, social cohesion and respect for life. Respect for people and for the earth is linked together in order for people to survive and care for future generations.[59]

This resonates with what Potawatomi scholar Kyle Whyte calls the concept of 'collective continuance': 'a society's overall adaptive capacity to maintain its members' cultural integrity, health, economic vitality, and political order into the future and avoid having its members experience preventable harms'.[60] For First Nations peoples, this is inextricably linked to their interdependent relationship with the natural world, what Bohaker refers to as 'kinship networks',[61] or what Kimmerer has termed a 'covenant of reciprocity' between humans and non-humans.[62] These reciprocal relationships of nurturing and support span across time in a way which implies duties to future members of the group.

The Aboriginal and Torres Strait Islander peoples of the continent of Australia and its coastal waters have philosophical and spiritual beliefs which centre on their relationship with their lands and which give rise to a strong sense of obligation towards the natural world, to ancestors and to descendants. For Aboriginal and Torres Strait Islander peoples, the relationship with their Country (both land and sea Country) is fundamental to their social and cultural systems. This goes far beyond just a recognition of the importance of the landscape as a provider of resources, and even beyond a sense of

58 Haudenosaunee Confederacy (n 54).

59 Clarkson, Morrissette and Ragallet (n 57) 7.

60 Kyle Powys Whyte, 'Food Sovereignty, Justice, and Indigenous Peoples' in Anne Barnhill, Mark Budolfson and Tyler Doggett (eds), *The Oxford Handbook of Food Ethics* (Oxford University Press, 2018) 345, 355; Kyle Whyte, 'Settler Colonialism, Ecology, and Environmental Injustice' (2018) 9(1) *Environment and Society* 125.

61 Heidi Bohaker, '"Nindoodemag": The Significance of Algonquian Kinship Networks in the Eastern Great Lakes Region, 1600–1701' (2006) 63(1) *The William and Mary Quarterly* 23.

62 Robin Wall Kimmerer, 'The Covenant of Reciprocity' in *The Wiley Blackwell Companion to Religion and Ecology* (Wiley, 2017) 368.

connection with the natural world. It is felt as a sense that people and nature are one and the same.

This understanding goes back to the creation stories of Aboriginal and Torres Strait Islander people, known in English as the Dreaming. The land was created by the movements of the ancestors, some of whom took animal form. The landscape and each animal within it has a story and a significance that is intimately related to the communities who live on those lands. As Indigenous scholar Mary Graham explains, 'every Aboriginal person has a part of the essence of one of the original creative spirits who formed the Australian landscape' and 'each human bears a creative and spiritual identity which still resides in land itself'.[63]

Bundjalung and Worimi Saltwater woman Phoebe McIlwraith explains: 'if the *gurulgam* (brolga crane) dance and we dance, if the *gagaru* (kookaburra) laugh and we laugh, then how could we as human beings ever see ourselves as inherently separate?' She explains that 'Indigenous peoples all around the world have independently drawn conclusions about how people do not "have a connection with nature", but that we are nature'.[64] Graham says: 'Aboriginal people's identity is essentially always embedded in land and defined by their relationships to it and to other people'.[65]

Consistent with this understanding, Country is more than just land or water. It was described by the twentieth-century anthropologist WEH Stanner as a 'sentient landscape'.[66] Both the landscape and the beings in it (both human and non-human) can feel. Gaagudju (Kakadu) elder Bill Neidjie explained the capacity for shared feelings between humans and the natural world: 'We brought up like all the animal and bird ... But all that animal for this world. Same us people'.[67] This is elaborated by Brigg and Graham: 'If humans listen carefully for these feelings and are open to them, then the feelings of Country and other beings can catch in our being and participate with it'.[68]

63 Mary Graham, 'Some Thoughts about the Philosophical Underpinnings of Aboriginal Worldviews – AHR' [2008] (45) *Australian Humanities Review* 2 <http://australianhumanitiesreview.org/2008/11/01/some-thoughts-about-the-philosophical-underpinnings-of-aboriginal-worldviews>.

64 Phoebe McIlwraith, 'Phoebe McIlwraith on Dancing with Country' [2022] *Groundswell* <www.groundswellgiving.org/journal/phoebe-mcilwraith>.

65 Graham (n 63) 5.

66 WEH Stanner, 'Aboriginal Territorial Organization: Estate, Range, Domain and Regime' (1965) 36(1) *Oceania* 1.

67 Bill Neidjie, *Story about Feeling* (Magabala Books, 1989) 14–15.

68 Morgan Brigg and Mary Graham, 'The Relevance of Aboriginal Political Concepts (5): Country, Place, and Territory', *ABC Religion & Ethics* (26 October 2020) <www.abc.net.au/religion/aboriginal-political-concepts-country-place-territory/12815608>.

The deep connection and shared identity between Aboriginal and Torres Strait Islander people and their lands and waters creates a sense of obligation to care for nature – as Graham says, 'the Land is the Law'.[69] Graham explains that the 'ancient, reciprocal relationship with nature' gives rise to 'an ethic of looking after, stewardship, caring for, and looking after the Land that nurtures'.[70] Graham calls this a custodial ethic, which is embodied in the practice of Caring for Country.[71] It has both a spiritual and practical purpose – practical because of the life-supporting role that nature plays for people, but also spiritual because of the sacredness of the land. McIlwraith explains the important role that human beings play in caring for land 'not as separate or superior beings, but as entities in kinship with and related to the non-human world'.[72]

The concept of Caring for Country has become embedded in environmental policy in much of Australia, where Indigenous knowledge is increasingly valued and Aboriginal and Torres Strait Islander people are more involved in leading environmental management practices. However, as Graham stresses, the custodial ethic inherent in Caring for Country is a philosophy, 'not just a solution for environmental management'.[73] Graham further explains that the custodial ethic requires people to look to the long term. 'It suggests a notion of caring for something outside ourselves, something that is in and of nature and that will exist for all time'.[74]

Flowing from the belief that people and land are one and the same, this custodial ethic establishes obligations not just towards Country but also towards other people. Graham says, 'all meaning comes from land … land determines our human-ness. Because land is sacred and must be looked after, the relationship between people and land becomes the template for society and social relations'.[75] The relationship with land extends to kinship *with the land* and all other living things and 'inculcates a sense of communal identity'.[76] Obligations and identity are therefore collective, relational and reflective, not individual, egoistic or survivalist.

These notions of custodianship towards land and obligation to kin exist across time as well as space. There are various nuanced concepts and words for

[69] Graham (n 63) 2.

[70] Mary Graham, *Custodial Navigator* (2013) 2 <www.indigenoussovereigntyaustralia.com.au/wp-content/uploads/2013/06/CustodialNavigator.pdf>.

[71] Graham (n 63); Graham (n 70).

[72] McIlwraith (n 64).

[73] Graham (n 63) 7; Brigg and Graham (n 68).

[74] Graham (n 63) 3.

[75] Ibid 2.

[76] Ibid 5.

time across the hundreds of groups and languages of First Nations peoples in Australia. They have in common a notion that time is not linear, or even cyclical, but is a deeper, relational understanding of the complex connections between people and the natural world. The Dreaming is not only past, it is also present as an ongoing, lived relationship with the stories of ancestors and Country. Those stories do not have clear beginnings and endings but instead 'happen continuously in circles and patterns'.[77] In other words, for Aboriginal and Torres Strait Islander people, time is 'always'.[78] The term 'everywhen', coined by WEH Stanner, attempts to capture this understanding of time, in which past, present and future coexist. He wrote: 'One cannot "fix" the Dreaming in time: it was, and is, everywhen'.[79]

The custodial relationship with Country therefore exists in the present as well as between ancestors and descendants. As McGrath and Rademaker explain, 'through a temporality that embraces continuity and a legacy for future generations, shared Indigenous narratives of place endorse the need for the land's ongoing maintenance and nurturing'.[80]

Indigenous knowledge systems such as those discussed here conceive obligations to future generations flowing from a deeper, relational understanding of humans' connection with nature. Such worldviews can be in stark contrast to the typically anthropocentric, exploitational and short-term view of many Western societies. As discussed later, for international human rights law to live up to its claims of universality, it too should accommodate a sense of obligation towards future people and the environment they will inherit.

2.3 INTERNATIONAL HUMAN RIGHTS LAW SHOULD ADVANCE INTERGENERATIONAL JUSTICE

It is one thing to say that current generations are morally obliged to have regard for future generations' interests. It is another to say that human rights law should be a place for this obligation to be performed. However, there are other reasons why human rights law should have more of a future focus and should work to advance intergenerational justice. These arguments align with Dryzek and Pickering's theory of ecological reflexivity, which calls on

77 Sara Kianga Judge, 'Starting Where You Are: First Nations Non-linear Storytelling', *The Australian Museum* <https://australian.museum/learn/first-nations/burra/non-linear-stories/australian.museum/learn/first-nations/burra/non-linear-stories>.

78 Jakelin Troy, Laura Rademaker and Ann McGrath, *Everywhen: Australia and the Language of Deep History* (NewSouth Publishing, 2023) 7; Kianga Judge (n 77).

79 WEH Stanner, *The Dreaming & Other Essays* (1956).

80 Troy, Rademaker and McGrath (n 78) 16.

institutions and governance frameworks to reflect on their performance and be capable of change in anticipation of ecological shifts.[81] Adapting this approach, human rights law needs to be able to evolve with changing global circumstances, particularly the triple planetary crises, but also a growing openness to other ideas about the role of human rights in that context.

2.3.1 *Human Rights Are Based on Equality*

Modern human rights law is said to have evolved from the theory of natural rights, which recognises human rights as those things that each person needs to live a life of dignity.[82] These rights are 'natural' because they are inherent in every human person. They have traditionally been thought of as minimum standards of treatment, safeguards of liberty and guarantees of basic needs, manifested as obligations owed by governments towards their citizens. Human rights law translates this understanding into a system of rights and obligations, possessed by rights-holders and duty-bearers who share a certain relationship.

Because human rights flow from our inherent human dignity, it follows that all human beings possess them equally: they are universal and are enjoyed without discrimination. The principle of equality is thus integral to human rights and is reflected in various provisions of international human rights law. The Universal Declaration of Human Rights begins in article 1 by declaring that 'all human beings are born free and equal in dignity and rights', and goes further in article 2 to explain that 'everyone is entitled to all the rights and freedoms set forth in this Declaration, without distinction of any kind'.[83]

The fundamental importance of equality and non-discrimination is carried into the two major human rights covenants. The International Covenant on Economic, Social and Cultural Rights (ICESCR) specifies in article 2 that all economic, social and cultural rights are to be enjoyed without discrimination:

> The States Parties to the present Covenant undertake to guarantee that the rights enunciated in the present Covenant will be exercised without discrimination of any kind as to race, colour, sex, language, religion, political or other opinion, national or social origin, property, birth or other status.[84]

81 Dryzek and Pickering (n 13).

82 David Boucher, *The Limits of Ethics in International Relations: Natural Law, Natural Rights, and Human Rights in Transition* (Oxford University Press, 2011) 3; see also Jack Donnelly, 'Human Rights as Natural Rights' (1982) 4(3) *Human Rights Quarterly* 391; Philip Alston, 'Making Space for New Human Rights: The Case of the Right to Development' (1988) 1 *Harvard Human Rights Yearbook* 3; John Finnis, *Natural Law and Natural Rights* (Oxford University Press, 2011).

83 *Universal Declaration of Human Rights* 1948 ('UDHR').

84 *International Covenant on Economic, Social and Cultural Rights* 1966 art 2(2) ('ICESCR').

The International Covenant on Civil and Political Rights (ICCPR) guarantees equality before the law in article 26:

> All persons are equal before the law and are entitled without any discrimination to the equal protection of the law. In this respect, the law shall prohibit any discrimination and guarantee to all persons equal and effective protection against discrimination on any ground such as race, colour, sex, language, religion, political or other opinion, national or social origin, property, birth or other status.[85]

While these provisions do not mention discrimination based on 'time of birth', if we accept that all humans enjoy human rights equally, the time at which a person is born should have no bearing on the rights they are entitled to. Indeed, the list of heads of discrimination has evolved over time, with disability, gender identity and sexual orientation all now regularly recognised. The fundamental connection between equality and human rights strongly suggests that human rights law should extend equally to future generations.

This was recognised in the Maastricht Principles, which begin by recognising that 'human rights extend to all members of the human family, including both present and future generations'. The principles rely on equality and dignity of all people as the basis on which to demand full and equal enjoyment of human rights for future generations.[86]

As noted, human rights law has its roots in Western, individualised ideas of identity. Historically, non-Western ideas have had less influence over and recognition within the body of international human rights law. However, human rights law is a living movement that is capable of evolution, and it should be open to the influence of other ways of thinking. For example, it took some time for IHRL to recognise collective rights, but this has become much more widely accepted. Now, jurisprudence and instruments like the United Nations Declaration on the Rights of Indigenous Peoples (UNDRIP) recognise that peoples possess collective rights 'which are indispensable for their existence, well-being and integral development as peoples'.[87] This has helped to bring the body of human rights law in line with non-Western, more relational understandings of human relations.

A similar process can (and should) apply so as to extend human rights protections temporally. As noted earlier, many cultures already recognise obligations towards their descendants, future people generally, and the

[85] *International Covenant on Civil and Political Rights* 1966 art 26 ('ICCPR').
[86] 'Maastricht Principles on the Human Rights of Future Generations' (n 2) 5.
[87] *United Nations Declaration on the Rights of Indigenous Peoples* 2007 ('UNDRIP').

natural world. If international human rights law is to live up to its claim of being based on equality and universality, and if it seeks to be a body of laws that speak to and reflect the experiences of all people, it should be open to these other ways of thinking. It should be able to accommodate more relational and collective obligations which are influenced by more fluid concepts of time and reflect a greater appreciation of inter-species and intergenerational connectivity. A commitment to equality therefore requires that we recognise that future generations possess rights and that current generations bear corresponding obligations.

2.3.2 *Human Rights Law Should Deliver Justice, including Intergenerational Justice*

In addition to its fundamental connection with equality, human rights law is also closely linked to justice. The foundational human rights instruments, the UDHR, ICCPR and ICESCR, all begin by stating that 'recognition of the inherent dignity and of the equal and inalienable rights of all members of the human family is the foundation of freedom, justice and peace in the world'.[88] One of the purposes of human rights law is therefore to contribute to justice, by safeguarding the rights through which justice is achieved. This is reinforced by the fact that IHRL, as a product of the United Nations and other multilateral bodies, inherits that institutional mandate to promote justice. The ability to achieve justice can be considered as critical to the legitimacy and effectiveness of international institutions, of which IHRL forms a part. Furthermore, human rights law is a chief framework for addressing injustice. When all these factors are taken into account, it is reasonable to expect that human rights law will help to achieve just outcomes.

As noted earlier, there are various theories of intergenerational justice, and different ideas about what current generations are obliged to do, but in general terms there is a strong basis for saying that justice requires current generations to act in a way which does not unfairly compromise the interests of future generations. This is particularly the case for environmental problems like climate change, which are recognised as a serious case of intergenerational injustice.[89] Because human rights law is linked at its core to ideas of justice

[88] *UDHR* (n 83); *ICCPR* (n 85); *ICESCR* (n 84) preambles.

[89] Henry Shue, 'Changing Images of Climate Change: Human Rights and Future Generations: Choosing a Future' (2014) 5 *Journal of Human Rights and the Environment* 50; Peter Lawrence, *Justice for Future Generations: Climate Change and International Law* (Edward Elgar, 2015); Achala Chandani, 'Distributive Justice and Sustainability as a Viable Foundation for the Future Climate Regime' (2007) 1(2) *Carbon & Climate Law Review* 12; Peter

and fairness, it should be attuned to future generations' rights and include mechanisms for ensuring these rights are respected, protected and fulfilled in relation to environmental and other threats.

2.3.3 *The Right to a Healthy Environment Incorporates Sustainability and Should Therefore Be Future Focused*

In recent developments at the United Nations, states have recognised the right to a clean, healthy and sustainable environment. The right was recognised by the Human Rights Council in 2021[90] and by the United Nations General Assembly in 2022.[91] While these resolutions are not legally binding, they demonstrate the increased receptiveness of states to environmental rights in general terms and reinforce the growing recognition of the right to a healthy environment within constitutional laws around the world.[92] And as Alston has explained, even without formal recognition as a legal right, the right to a healthy environment has growing normative importance.[93] These developments are discussed in more detail in Chapter 3.

This integration of sustainability and human rights implies a future-focused dimension to environmental rights. Sustainable development has been defined as 'development that meets the needs of the present without compromising the ability of future generations to meet their own needs'.[94] While sustainable development does not explicitly adopt the language of human rights, it is understood as a guiding principle which shares common values and a mutually supportive relationship with human rights. It is also closely linked with theories of intergenerational equity. As Edith Brown Weiss says, 'sustainable development relies on a commitment to equity with future generations'.[95] By recognising the right to a healthy environment, human

Lawrence, 'Justifying Representation of Future Generations and Nature: Contradictory or Mutually Supporting Values?' (2022) 11(3) *Transnational Environmental Law* 553.

90 Human Rights Council, 'The Human Right to a Safe, Clean, Healthy and Sustainable Environment' <https://documents-dds-ny.un.org/doc/UNDOC/LTD/G21/270/15/PDF/G2127015.pdf?OpenElement> ('Resolution 48/13').

91 United Nations General Assembly (n 1) (161 votes in favour, nil against and 8 abstentions).

92 Philip Alston, 'The Right to a Healthy Environment: Beyond Twentieth Century Conceptions of Rights' (2023) 117 *AJIL Unbound* 167.

93 Ibid.

94 World Commission on Environment and Development, *Our Common Future: Report of the World Commission on Environment and Development* (No A/42/427, 1987) <www.un-documents.net/wced-ocf.htm>; United Nations Environment Programme, *Rio Declaration on Environment and Development* (No A/Conf.151/26, 14 June 1992).

95 Brown Weiss, 'In Fairness to Future Generations and Sustainable Development' (n 15) 19.

rights law incorporates these ideas of sustainability and, by extension, intergenerational equity.

Together, the connections between human rights law and justice, equality and sustainability all suggest that IHRL is an appropriate forum to advance the rights of future generations. Indeed, these correlations suggest that the rules, processes and institutions of human rights law should adapt to better protect future generations' rights.

2.4 CONCLUSION

This chapter has attempted to justify the central premise of the book: international human rights law needs to adapt to recognise and protect the environmental rights of future generations. Theories of intergenerational justice and equity support the notion that current generations should be taking steps to protect future generations' interests, something which is well understood within the worldviews of many Indigenous peoples. International human rights law is an appropriate place for these obligations to be embedded, since human rights law is based on concepts of justice and equality. And there is no good reason to believe that human rights should not be extended to future people, since we know that they will have dignity and interests equal to those of current generations.

Drawing those threads together, it is clear not only that human rights law can in theory accommodate future generations, but also that we should work to make this happen. The recognition of the right to a healthy environment has gone some way to achieving this, through the integration of sustainability. However, as the next two chapters will show, this right has only received limited protection at the international level, and the rules and procedures of international human rights law so far make it impossible for future generations' rights to be meaningfully protected.

3

The Law

What the Existing Legal Framework Says about the Environmental Rights of Future Generations

3.1 INTRODUCTION

The central purpose of this book is to look for ways to enhance the international protection of environmental rights so that they can better serve the interests and well-being of future generations. Before that task can begin, it is necessary to take stock of the current state of international law in relation to both environmental rights and the interests of future generations. As this chapter will show, these laws are at best a patchwork, albeit one whose seams have been reinforced by a number of key developments over recent years. The assorted pieces consist of international and regional human rights treaties and soft law instruments, interpreted by courts and tribunals and influenced by domestic laws.

The chapter begins with a discussion of one of the most significant events in the history of environmental rights: the adoption by the United Nations General Assembly of a resolution recognising the right to a clean, healthy and sustainable environment.[1] This resolution was the culmination of a sustained body of work by Special Rapporteurs and other UN agencies, civil society and a small number of states who championed the cause. The resolution recognises the many fundamental ways in which human rights and the environment are interdependent and brings attention to the urgent need to

[1] United Nations General Assembly, 'Resolution 76/300: The Human Right to a Clean, Healthy and Sustainable Environment'.

address environmental degradation that threatens the lives of millions of people around the world.

The resolution itself is not legally binding, however, and does not create any new obligations for states, although it could influence the interpretation of other laws. Following a discussion of the resolution, the chapter will outline those other laws, beginning with an overview of relevant international and regional human rights laws which are considered part of the body of environmental rights. It will also consider key components of international environmental law which have been used to interpret human rights obligations and which could help to address intergenerational issues. Finally, the chapter identifies some significant domestic instruments and mechanisms which seek to safeguard future generations' rights and considers the extent to which these might influence laws at the international level.

The overall picture painted in this chapter is of a body of international law which recognises the mutually supportive relationship of human rights and environmental law, but which falls short of creating binding obligations on states in relation to that broader relationship. Where human rights are recognised to have environmental dimensions, these are typically defined rather narrowly, such that environmental degradation is actionable where it causes direct, actual or imminent harm to the enjoyment of human rights. Environmental or ecological well-being generally is not recognised as a human right and future environmental harm largely falls outside the scope of existing law. The scope and content of existing environmental rights are therefore difficult to utilise in the pursuit of future generations' rights. Further, as Chapter 4 will elaborate, the procedures and principles of international law make it very difficult to bring claims on behalf of future generations, limiting the opportunities for courts and tribunals to expand the scope of existing laws through interpretation. Together, this chapter and the next lay the foundation for the proposed changes set out in Chapter 5.

3.2 ENVIRONMENTAL RIGHTS IN INTERNATIONAL HUMAN RIGHTS LAW

In this book 'environmental rights' are defined broadly to include both standalone rights to a healthy environment or similar dedicated rights, as well as the environmental dimensions of other human rights like the right to life or the right to an adequate standard of living. As this section will show, at the international level environmental rights are made up almost entirely of the latter category, in which the environment is viewed as a necessary precondition to the enjoyment of other rights and influences the content of those rights and the obligations that are associated with them. Before turning to those rights, however, this section first examines the high point of the right to a healthy environment in international law: the adoption by the General Assembly of Resolution 76/300.

3.2.1 *The Right to a Healthy Environment*

In July 2022, the United Nations General Assembly adopted Resolution 76/300 on the right to a clean, healthy and sustainable environment.[2] It followed the adoption the previous year of a similar resolution by the Human Rights Council.[3] The General Assembly resolution was introduced by Costa Rica and was passed by an astounding majority, with 161 states voting in favour and none against, with only 8 states abstaining.

The resolution builds on the considerable body of work contributed by successive Special Rapporteurs on human rights and the environment, John Knox and David Boyd, who recommended that the General Assembly recognise the right to a clean, healthy and sustainable environment.[4] The text of the resolution does not define the right to a clean, healthy and sustainable environment, but simply declares that it is recognised as a right.[5] The preamble notes the various ways that protection of the environment is important for the realisation of human rights and that environmental degradation has negative implications for the enjoyment of rights. It notes the special vulnerability of women, children, people with disabilities, Indigenous peoples and other groups whose rights can be particularly threatened through environmental degradation. The brief operative portion of the resolution also notes that 'the right to a clean, healthy and sustainable environment is related to other rights and existing international law'.[6]

2 Ibid.

3 United Nations Human Rights Council, 'Resolution 48/13: The Human Right to a Clean, Healthy and Sustainable Environment' <https://documents-dds-ny.un.org/doc/UNDOC/GEN/G21/289/50/PDF/G2128950.pdf?OpenElement>.

4 John Knox and David Boyd, *Report of the Special Rapporteur on the Issue of Human Rights Obligations Relating to the Enjoyment of a Safe, Clean, Healthy and Sustainable Environment* (United Nations General Assembly, 19 July 2018) <http://undocs.org/A/73/188>; David Boyd, John Knox and Marc Limon, *The Time Is Now: The Case for Universal Recognition of the Right to a Safe, Clean, Healthy and Sustainable Environment* (Universal Rights Group, February 2021); John H Knox, *Framework Principles on Human Rights and the Environment* (No A/HRC/37/59 (annex), 2018) <www.ohchr.org/Documents/Issues/Environment/SREnvironment/FrameworkPrinciplesUserFriendlyVersion.pdf>.

5 United Nations General Assembly (n 1) [1].

6 Ibid [2].

Taken as a whole, the resolution might best be characterised as framing the right to a clean, healthy and sustainable environment as a 'synthesis' right[7] which captures the many ways in which the environment is necessary for the enjoyment of human rights, and vice versa. It does not recognise a standalone right to an environment of a particular quality, but instead articulates the essential interdependence of human rights and the environment. In doing so the resolution does not create any new obligations for states but rather interprets existing duties – no doubt a key reason why it attracted such strong agreement among voting UN members.

A related development within the UN system, though one which attracted less public attention, is the recent General Comment of the Committee on the Rights of the Child. In General Comment 26 on Children's Rights and the Environment, the Committee refers to a clean, healthy and sustainable environment 'as both a right itself and necessary for the fulfilment of a broad range of children's rights'.[8] The General Comment is a more detailed document which has benefited from extensive consultation with civil society and children themselves (certainly an advantage afforded by the fact that General Comments do not require approval by voting states). It explains that the right to a healthy environment is 'implicit' in the Convention on the Rights of the Child and directly linked to the rights to life, survival and development, to the highest attainable standard of health, to an adequate standard of living, and to education, which includes the development of respect for the natural environment.[9]

Unlike Resolution 76/300, General Comment 26 devotes some time to expanding on the content of the right to a healthy environment, explaining that the substantive elements include 'clean air, a safe and stable climate, healthy ecosystems and biodiversity, safe and sufficient water, healthy and sustainable food and non-toxic environments'.[10] It also sets out a number of actions which states should take to implement the right and addresses procedural components such as ensuring children have access to information and can participate in decision-making.[11] General Comment 26 represents a

7 Philippe Cullet, 'Definition of an Environmental Right in a Human Rights Context' (1995) 13(1) *Netherlands Quarterly of Human Rights* 25; see also Luis Rodriguez-Rivera, 'Is the Human Right to Environment Recognised under International Law? It Depends on the Source' (2001) 12 *Colorado Journal of International Environmental Law and Policy* 1.

8 Committee on the Rights of the Child, 'General Comment No. 26 on Children's Rights and the Environment with a Special Focus on Climate Change' [8].

9 Ibid [63].

10 Ibid [64].

11 Ibid [65].

significant development in international recognition of the right to a clean, healthy and sustainable environment. It is arguably the most comprehensive attempt by an international human rights body to flesh out the content of such a right beyond merely restating the environmental dimensions of other rights.

In earlier work, I have argued that the international community would not need to recognise an independent right to a healthy environment in international human rights law as long as we properly recognise and protect the environmental dimensions of existing rights.[12] Despite my earlier reservations, General Comment 26 and General Assembly Resolution 76/300 are very welcome developments. Together they confirm the importance of the environment for the enjoyment of human rights, and Resolution 76/300 also represents a strong statement of support from states. While it is hoped that they will be influential in the interpretation of human rights treaties, they do not themselves expand the scope of those laws.[13] It remains necessary to explore the ways in which other human rights law encompasses the environment.

3.2.2 *The Right to Life*

The right to life has been raised in several environmental rights cases, usually where environmental conditions present threats to life in the short or medium term.[14] State obligations under the right to life entail not just a negative obligation not to take life, but also a positive obligation to protect against threats to life. Further, the right to life has been interpreted to go beyond just a right to 'any life' but to entail a right to a life of dignity that is worthy of a human being.[15]

The Human Rights Committee has explained that the right to life protected under article 6 of the International Covenant on Civil and Political Rights (ICCPR) requires a certain quality of environment, and it has declared that it considers climate change a serious threat to the right to life.[16] However, it has been somewhat reluctant to find breaches of the right to life

[12] Bridget Lewis, *Environmental Human Rights and Climate Change: Current Status and Future Prospects* (Springer, 2018).

[13] For a useful analysis of the adoption and significance of the GA resolution, see Philip Alston, 'The Right to a Healthy Environment: Beyond Twentieth Century Conceptions of Rights' (2023) 117 *AJIL Unbound* 167.

[14] See e.g. *Budayeva and Others v Russia* [2008] Eur Ct HR, Application Nos 15349/02, 21166/02, 20058/02, 11673/02 and 15343/02.

[15] Human Rights Committee, *General Comment No. 36: Article 6 (the Right to Life)* (30 October 2018).

[16] Ibid [260].

in the two climate change cases that have come before the Committee to date. In *Teitiota v New Zealand*, the Committee concluded that the right to life could be threatened by climate change in Mr Teitiota's home country of Kiribati, but the threat was not yet imminent enough to trigger New Zealand's positive obligations to protect him from that harm. The Committee was of the view that rising sea levels would not cause significant interference with human rights for ten to fifteen years, which allowed time for steps to be taken to address the risk.[17] In *Billy v Australia*, the Committee found violations of other rights in the ICCPR and therefore felt it unnecessary to consider the claim based on article 6.[18]

The European Court of Human Rights has developed a test of 'real and immediate risk' with respect to the right to life found in article 2 of the European Convention on Human Rights (ECHR). Under this approach, a threat to life must be both foreseeable and imminent before a state will have positive duties of protection.[19] However, despite an emphasis on immediacy in some cases, the court has been prepared to find a violation based on the general foreseeability of a threat to life, even where it may not materialise for some time.[20]

The Inter-American Court of Human Rights has followed a similar approach, requiring a real and imminent danger as the threshold for state duties of protection to crystallise. As in Europe, interpretations of 'imminent' have varied. In some cases, the court has accepted that a general knowledge of circumstances which present a risk to life is enough to trigger state duties.[21] In others, the court has said that a specific threat to individual people was required before protection was required.[22]

The Inter-American human rights bodies have also given detailed consideration to the implications of the right to life for Indigenous peoples. As early as

[17] *Ioane Teitiota v New Zealand* [2019] Human Rights Committee Communication No 2728/2016.

[18] *Daniel Billy and Others v Australia* [2022] Human Rights Committee Communication No 3624/2019.

[19] *Osman v the United Kingdom* [1998] Eur Court HR 23452/94; Justine Bell-James and Briana Collins, 'Human Rights and Climate Change Litigation: Should Temporal Imminence Form Part of Positive Rights Obligations?' (2022) 13(1) *Journal of Human Rights and the Environment* 212.

[20] *Öneryildiz v Turkey* [2004] Eur Court HR [GC] 27785/10.

[21] See e.g. *Case of the Pueblo Bello Massacre v Colombia* (Inter-American Court of Human Rights, 31 January 2006); *Valle Jaramillo et al v Colombia (Merits, Reparations, and Costs)* [2008] Ser C No 192 Inter-Am Court HR Case No 12.415.

[22] *González et al ('Cotton Field') v Mexico (Preliminary Objection, Merits, Reparations, and Costs)* [2009] Ser C 205; *Jessica Lenahan (Gonzales) v United States* [2011] Report No 80/11 (2011) Inter-American Court of Human Rights Case No 12.626.

1985 the Inter-American Court of Human Rights recognised in the *Yanomami* case that the displacement of Indigenous owners from their land to facilitate the construction of a road had such a significant impact on their well-being that it constituted a breach of the right to life.[23] Likewise, in *Yakye Axa*, an Indigenous group who had been moved off their lands to facilitate a development project lost access to their means of subsistence, which was considered to constitute a breach of their right to life.[24]

3.2.3 *The Right to Private and Family Life*

The ICCPR guarantees freedom from arbitrary or unlawful interference with privacy, family, home and correspondence.[25] The right is included, albeit in slightly different language, in all regional human rights treaties and most national laws. It is commonly referred to as the right to privacy, but it is the protection of family and home life which is more frequently invoked in environmental cases. In fact, in terms of environmental rights case law, the right to private and family life has received more consideration by international and regional judicial bodies than any other right. The principal forum for this work has been the European Court of Human Rights and as a result we have a sizable volume of commentary on the relationship between the right to private and family life (protected in article 8 of the ECHR) and the environment.

While the European Court has stressed that there is no human right to a good environment *per se*,[26] environmental harm can interfere with the enjoyment of private and family life to such a degree that it constitutes a violation of the Convention. For example, the failure of government to properly manage pollution and other environmental hazards from industrial activities, waste treatment plants and mining facilities has been recognised as a violation of article 8.[27]

23 *Comunidad Yanomami Caso, No 7615, Resolución No 12/85 [ENG]* (Inter-American Commission on Human Rights, 5 March 1985).

24 *Yakye Axa Indigenous Community v Paraguay (Merits, Reparations and Costs)* (17 June 2005) 560.

25 *International Covenant on Civil and Political Rights* 1966 art 17 ('ICCPR').

26 *Kyrtatos v Greece* [2003] Eur Court HR 41666/98.

27 *Lopez Ostra v Spain* [1994] Eur Court HR 16798/90; *Dubetska v Ukraine* [2011] Eur Court HR 30499/03; *Fadeyeva v Russia* [2005] Eur Court HR, Application No 55723/00; *Taskin v Turkey* (Eur Court HR, 2004); *Giacomelli v Italy* [2006] Eur Court HR, Application No 59909/00; *Cordella and Others v Italy* [2019] Eur Court HR, Application Nos 54414/13 and 54264/15.

For a long time, article 8 was only successfully applied in relation to quite localised impacts where the impact on individual lives was quite pronounced.[28] However, in the more recent case of *Verein KlimaSeniorinnen Schweiz v Switzerland*, the European Court recognised that climate change could impact on the right to private and family life and, where governments fail to take adequate action to prevent that harm, they can be in breach of their Convention obligations.[29] The United Nations Human Rights Committee has made a similar finding in *Daniel Billy v Australia*, where the effects of climate change on the complainants' home islands in the Torres Strait constituted a breach of the ICCPR.[30]

3.2.4 *The Right to an Adequate Standard of Living*

The right to an adequate standard of living encompasses rights to adequate food, water and housing.[31] It has obvious environmental preconditions and as such is susceptible to interference through many forms of environmental degradation, pollution or destruction. The Committee on Economic, Social and Cultural Rights has issued a number of general comments that address different aspects of the right to an adequate standard of living. In relation to the right to adequate food, it has explained that states must adopt appropriate environmental policies to ensure the availability of adequate food supplies, which also entails the sustainable management of natural resources.[32] Similarly, the right to water entails several obligations to protect and maintain environmental health, including preventing water pollution and reducing the potential for aquatic ecosystems to provide habitat for disease vectors.[33]

The right to water is also mentioned in the Convention on the Elimination of Discrimination against Women, which specifically notes the need to reduce barriers for women in rural areas to access adequate clean water supplies.[34] The Convention on the Rights of the Child includes access to nutritious foods and clean drinking water in the right of children to the highest attainable standard of

28 Elinor Buys and Bridget Lewis, 'Environmental Protection through European and African Human Rights Frameworks' (2022) 26(6) *International Journal of Human Rights* 949.

29 *Verein KlimaSeniorinnen Schweiz v Switzerland* [2024] Eur Court HR, Application No 53600/20.

30 *Daniel Billy and Others v Australia* (n 18).

31 *International Covenant on Economic, Social and Cultural Rights* 1966 art 11 ('ICESCR').

32 Committee on Economic, Social and Cultural Rights, *General Comment No. 12: The Right to Adequate Food (Art. 11)* (1999) [14], [25].

33 Committee on Economic, Social and Cultural Rights, *General Comment No. 15: The Right to Water* (2003) [8].

34 *Convention on the Elimination of All Forms of Discrimination against Women* 1979 art 14 (2)(h).

health.[35] There are obviously strong links between the rights to food and water and the right to health, discussed later.

There are very few environmental rights cases that have been based on the right to an adequate standard of living, however. This seems to be because, where environmental conditions have impacted on the standard of living, such circumstances will usually also give rise to a breach of other rights such as the right to health or the right to private or family life or, if serious enough, the right to life itself.

3.2.5 *The Right to the Highest Attainable Standard of Health*

There is much that can be said about the environmental dimensions of the right to health, since the connections between environmental conditions and health are many and varied. The right to the highest attainable standard of health is recognised in the International Covenant on Economic, Social and Cultural Rights (ICESCR) in article 12, which notes that the 'improvement of environmental hygiene' is an important part of achieving this right.[36] The *Convention on the Rights of the Child* guarantees this right for children and requires states to consider the 'dangers of environmental pollution'.[37] The general comments of both the Committee on the Rights of the Child and the Committee on Economic, Social and Cultural Rights explain the various environmental factors which are necessary to the enjoyment of this right.[38] These include ensuring adequate sanitation, potable water and nutritious foods as well as combatting soil, water and air pollution.

The European Convention does not include a right to health and the Committee on Economic, Social and Cultural Rights has only had the power to hear individual complaints since 2008.[39] As a result, there are comparatively fewer cases examining the right to health. The case of *Sacchi v Argentina* raised arguments in relation to the right to health before the Committee on the Rights of the Child; however, as discussed elsewhere, this case did not proceed to a hearing of the merits.[40] The African and American

35 *Convention on the Rights of the Child* 1989 art 24 (2)(c) ('*CRC*').

36 *ICESCR* (n 31) art 12 (2)(b).

37 *CRC* (n 35) art 24 (2)(c).

38 Committee on the Rights of the Child (n 8) 26; Committee on Economic, Social and Cultural Rights, *General Comment No. 14: The Right to the Highest Attainable Standard of Health* (2000) 14.

39 *Optional Protocol of the Covenant on Economic, Social and Cultural Rights* 2008.

40 *Sacchi et al v Argentina* [2021] Committee on the Rights of the Child Communication No 104/2019.

regional human rights systems have heard cases relating to the right to health.[41] The latter especially has noted the potential for environmental degradation to impact negatively on the physical, emotional and spiritual health of Indigenous people.[42]

3.2.6 *The Rights of Indigenous Peoples*

The rights of future generations and environmental rights are frequently discussed in the context of the rights of Indigenous peoples. These are articulated in the United Nations Declaration on the Rights of Indigenous Peoples (UNDRIP), which builds on human rights treaties and elaborates states' obligations with respect to the particular rights of Indigenous peoples.

The right to self-determination is a cornerstone of Indigenous rights, and it has a long history within international human rights law. The self-determination and independence of colonised peoples was an objective of the fledgling United Nations when it formed after the Second World War, and the recognition of self-determination as a fundamental right is reflected in its inclusion in common article 1 of the ICCPR and ICESCR.[43] That provision recognises that all peoples have the right 'to freely determine their political status and freely pursue their economic, social and cultural development'. UNDRIP articulates the many connections between the environment and self-determination, not least of all the right of Indigenous peoples to control and benefit from their lands, natural resources and traditional knowledge.[44]

UNDRIP also draws legal force from article 27 of the ICCPR, which guarantees the rights of members of minority groups 'in community with the other members of their group, to enjoy their own culture, to profess and practise their own religion, or to use their own language'. Article 27 has been interpreted by the Human Rights Committee to protect traditional activities such as fishing and hunting,[45] as well as the right to live in particular areas designated by law.[46] In its recent decision in *Billy v Australia*, the

[41] *Social and Economic Rights Action Center and the Center for Economic and Social Rights v Nigeria* [2002] African Commission on Human and Peoples' Rights No 155/96 ('*Ogoniland Decision*').

[42] *Yakye Axa Indigenous Community v Paraguay (Merits, Reparations and Costs)* (n 24).

[43] *Charter of the United Nations* 1945 art 1; *ICCPR* (n 25) art 1; *ICESCR* (n 31) art 1.

[44] *United Nations Declaration on the Rights of Indigenous Peoples* 2007 arts 25–29, 31 ('*UNDRIP*').

[45] *Ivan Kitok v Sweden* [1988] Human Rights Committee Communication No 197/1985.

[46] *Ominayak (Lubicon Lake Band) v Canada* [1990] Human Rights Committee Communication No 167/1984.

Committee found that Australia's failure to protect Torres Strait Islanders from the impacts of climate change amounted to a failure to protect their 'collective ability to maintain their traditional way of life, to transmit to their children and future generations their culture and traditions and use of land and sea resources'. This constituted a violation of article 27.[47] These interpretations of the rights to culture and self-determination emphasise the ways in which environmental protection is a necessary precondition for the full enjoyment of rights, but the reverse is also true. Meaningful self-determination exercised by Indigenous people has been shown to have profound benefits for the environment.[48]

These cases also show the importance of recognising the collective dimension of environmental rights. The right to self-determination is recognised as a collective right, enjoyed and exercised by a people together as a group. Cultural rights are also generally viewed as group rights. Although it is possible for the right to culture to be violated or enforced on an individual basis, the full enjoyment of the right is something that requires collective expression. Over the years, human rights scholars have debated about whether – and which – human rights can be said to be group rights. Some scholars, like Donnelly and Cranston, expressed a view that collective rights found in international human rights law, most obviously the right of peoples to self-determination, are distinct from human rights proper.[49] Others, like Crawford, Brownlie and Alston, were supportive of collective rights, particularly where recognition of group rights is necessary to adequately protect the interests of ethnic and religious minorities or other communities.[50]

It is fair to say, I think, that these debates about whether human rights can be possessed by groups or communities have been put to rest by UNDRIP and the many constitutional reforms around the world that have recognised the rights of Indigenous communities, minorities and communities. Cultural practices, languages, knowledge systems and spiritual beliefs are collective

47 *Daniel Billy and Others v Australia* (n 18) [8.14].

48 Teagan Goolmeer et al, 'Recognizing Culturally Significant Species and Indigenous-Led Management Is Key to Meeting International Biodiversity Obligations' (2022) 15(6) *Conservation Letters* e12899; Graeme Reed et al, 'Indigenous Guardians as an Emerging Approach to Indigenous Environmental Governance' (2021) 35(1) *Conservation Biology* 179.

49 Jack Donnelly, 'In Search of the Unicorn. The Jurisprudence and Politics of the Right to Development' (1985) 15(3) *California Western International Law Journal* 474; Maurice Cranston, *What Are Human Rights?* (Taplinger, 1973).

50 James Crawford (ed), *The Rights of Peoples* (Clarendon Press, 1988); Ian Brownlie, 'The Rights of Peoples in Modern International Law' (1985) 9 *Bulletin of the Australian Society of Legal Philosophy* 104; Philip Alston, 'The Shortcomings of a "Garfield the Cat" Approach to the Right to Development' (1985) 15(3) *California Western International Law Journal* 510.

goods which must be protected for the good of the group as well as its individual members. Similarly, our relationship to the environment also has a collective dimension, as it is something that we experience and benefit from in community with other members of society (or other members of the ecosystem).

The Inter-American Court of Human Rights recognised this shared experience of environmental rights in its landmark Advisory Opinion in 2017. The court was requested by Colombia to provide advice to states on their obligations with respect to the environment under the American Convention on Human Rights. The court confirmed the 'undeniable relationship between the protection of the environment and the realization of other human rights'.[51] It also stressed the importance of protecting Indigenous peoples' collective ownership of their lands and shared pursuit of their common interests. The most significant finding of the court, however, was its recognition of a standalone right to a healthy environment.

The court observed that environmental degradation can cause irreparable damage to human beings, and for this reason, a healthy environment is a fundamental right for the existence of humankind. In explaining the nature of this right, the court stressed that it is a right with both individual and collective dimensions. As an individual right, it operates similarly to other human rights, like the rights to life, health and personal integrity, which import individualised obligations to the state to safeguard quality of life. In its collective dimension, 'the right constitutes a universal value that is owed to present and future generations'.[52] This finding has been recognised as a critical milestone in the recognition of both environmental rights and the rights of future generations.[53]

The court went on to articulate the obligations that flow from the right to a healthy environment, noting the value of international environmental law in informing the nature of states' duties. As such, the obligations that attach to the right to a healthy environment include an obligation to prevent significant

[51] *Environment and Human Rights: State Obligations in Relation to the Environment in the Context of the Protection and Guarantee of the Rights to Life and to Personal Integrity (Advisory Opinion)* [2017] Inter-American Court of Human Rights OC-23/17, 45.

[52] Ibid [59].

[53] Christopher Campbell-Duruflé and Sumudu Anopama Atapattu, 'The Inter-American Court's Environment and Human Rights Advisory Opinion: Implications for International Climate Law' (2018) 8(3–4) *Climate Law* 321; Angeliki Papantoniou, 'Advisory Opinion on the Environment and Human Rights' (2018) 112(3) *American Society of International Law* 460; Maria Antonia Tigre and Natalia Urzola, 'The 2017 Inter-American Court's Advisory Opinion: Changing the Paradigm for International Environmental Law in the Anthropocene' (2021) 12(1) *Journal of Human Rights and the Environment* 24.

environmental damage and adhere to the precautionary principle, to follow procedural rules relating to environmental impact assessments, public participation and access to justice, and to cooperate in good faith with other states. The court did not address the principle of intergenerational equity specifically, though its influence can be detected in the discussion of sustainable development. The following discussion expands on the way that sustainable development and international environmental law more broadly promote intergenerational equity.

3.3 INTERGENERATIONAL EQUITY IN INTERNATIONAL LAW

Intergenerational equity is considered one of the key principles of international environmental law. It can be found in a number of influential instruments and is closely related to the other principles of sustainable development and the precautionary principle. As Judge Cançado Trindade said in his separate opinion in *Pulp Mills on the River Uruguay*: 'Nowadays, in 2010, it can hardly be doubted that the acknowledgement of intergenerational equity forms part of conventional wisdom in international environmental law'.[54] However, while it is an important principle, it is not in itself legally binding and in many of the treaties in which it appears it has a more recommendatory than mandatory character.[55] The following discussion illustrates this through a brief look at international environmental law, the climate change regime and recent soft law.

References to future generations or the principle of intergenerational equity appear in many environmental treaties. For example, the *Convention on Biological Diversity* mentions future generations as part of the definition of sustainable use, which is one of the objectives of the convention.[56] The duty of states to conserve and transmit cultural and natural heritage to future generations is a key obligation under UNESCO's *World Heritage Convention*.[57] And where natural resources must be managed according to the principle of sustainable development, such management is also implicitly done for the benefit of future generations. Sustainable development is defined as 'development that meets the needs of the present without compromising

[54] *Case Concerning Pulp Mills on the River Uruguay (Argentina v Uruguay)* [2010] ICJ Rep 14 (Separate Opinion Judge Cançado Trindade) 181.

[55] Lydia Slobodian, 'Defending the Future: Intergenerational Equity in Climate Litigation' (2020) 32 *Georgetown Environmental Law Review* 569.

[56] *Convention on Biological Diversity* 1992 arts 1–2.

[57] *Convention Concerning the Protection of the World Cultural and Natural Heritage* 1972.

the ability of future generations to meet their own needs'.[58] However, as Slobodian says, 'though there is international agreement on a general need to consider the interests of future generations in environmental decision making, there is no certainty on what this entails or what rights, if any, it implies'.[59]

International climate change law includes intergenerational equity as a principle that should guide states' policies, but falls short of creating a binding expectation that could be enforced on behalf of future generations. The foundation of the regime is the *United Nations Framework Convention on Climate Change* (UNFCCC), which sets out several guiding principles that states should follow. Among these is the provision that 'the Parties should protect the climate system for the benefit of present and future generations of humankind'.[60]

When states were negotiating the Paris Agreement in 2015, they had the opportunity to build on this guiding principle and do more to protect the rights of future generations against climate change. Instead, they arguably regressed even from the mild language of the UNFCCC. Earlier drafts of the Paris text had included an option to list action 'for the benefit of present and future generations' among the objectives in article 2.[61] Instead, this was removed, and the final text includes no explicit reference to future generations.

The Preamble refers to intergenerational equity among many principles which states should respect:

> Acknowledging that climate change is a common concern of humankind, Parties should, when taking action to address climate change, respect, promote and consider their respective obligations on human rights, the right to health, the rights of indigenous peoples, local communities, migrants, children, persons with disabilities and people in vulnerable situations and the right to development, as well as gender equality, empowerment of women and intergenerational equity.[62]

The inclusion of human rights in this provision is also noteworthy, since it was the first time human rights had appeared in an instrument of the global

58 Report of the World Commission on Environment and Development, Our Common Future, UN Doc A/42/427, 4 August 1987; United Nations Environment Programme, Rio Declaration on Environment and Development, UN Doc A/Conf.151/26, 14 June 1992.

59 Slobodian (n 55) 573.

60 *United Nations Framework Convention on Climate Change* 1992 art 3 ('UNFCCC').

61 Ad Hoc Working Group on the Durban Platform for Enhanced Action, 'Negotiating Text, 2nd Sess, Part 8, Geneva, Switzerland' Section C, art 2.2.

62 *Paris Agreement* 2015 Preamble, para 11.

climate regime. However, these preambular statements are not in the operative part of the Agreement and are not therefore legally binding.[63]

The principle of sustainable development has a more prominent place in the Paris Agreement, appearing in several operative provisions including the guiding principles listed in article 2. It is therefore arguable that states have at least an indirect obligation to address future generations' rights in devising their climate policies. Overall, though, the global climate regime does little more than acknowledge that future generations have a stake in our current climate actions. States have been reluctant to translate this recognition into any legally binding obligation, and as a result the regime does not reflect the seriousness of the risks facing future generations.

Many of the more detailed and compelling references to future generations and intergenerational equity appear in soft law instruments dating back several decades. The 1972 Stockholm Declaration on the Human Environment begins in Principle 1 by declaring states' 'solemn responsibility to protect and improve the environment for present and future generations'.[64] Twenty years later, the Rio Declaration on Environment and Development articulated the modern notion of sustainable development in stressing that 'the right to development must be fulfilled so as to equitably meet developmental and environmental needs of present and future generations'.[65] In 2012, the commitment to sustainable development was yet again restated at the Rio+20 summit in an outcome document titled 'The Future We Want'.[66] Despite these periodic declarations of intergenerational solidarity, international environmental law has done little to translate intergenerational equity into a legal norm.

At the Rio+20 summit, states requested the United Nations Secretary-General, at the time Ban Ki-moon, to report back on 'the need for promoting intergenerational solidarity for the achievement of sustainable development, taking into account the needs of future generations'.[67] The resulting report, entitled *Intergenerational Solidarity and the Needs of Future Generations*, canvassed a number of options for improving

[63] For more discussion, see Bridget Lewis, 'The Rights of Future Generations within the Post-Paris Climate Regime' (2018) 7(1) *Transnational Environmental Law* 69.

[64] *Declaration of the United Nations Conference on the Human Environment (Stockholm Declaration)* 1972.

[65] United Nations Environment Programme, *Rio Declaration on Environment and Development* (No A/Conf.151/26, 14 June 1992).

[66] Rio+20 Conference on Sustainable Development, *The Future We Want (Outcome Document)* (2012) <https://sustainabledevelopment.un.org/rio20/futurewewant>.

[67] Ibid [86].

institutional arrangements within the UN, but did not propose a specific recommendation.[68]

The current Secretary-General, António Guterres, has moved this project forward considerably through a report and strategy titled *Our Common Agenda*. The strategy includes a number of actions, including the appointment of a Special Envoy for Future Generations.[69] The Special Envoy would be tasked with awareness-raising and advocating for the interests of future generations, in a similar fashion to the role performed by Special Rapporteurs and other thematic mandate-holders within the area of human rights.

The Secretary-General also suggested that states adopt a Declaration on Future Generations, to consolidate existing commitments in national laws and signal an international intention to address long-term challenges. To advance this proposal, a Summit of the Future was convened in September 2024, at which national delegates negotiated the final text of the Declaration.[70] The Declaration is annexed to the Pact for the Future, a larger document which sets out a range of actions that states committed to take in relation to peace and security, sustainable development, technology, and governance.[71] The Declaration then expands on these commitments and sets out a range of actions aimed at addressing today's structural inequities, poverty, conflict, and environmental degradation, as well as strengthening international cooperation and advancing technology, in order to leave a better world for future generations.

However, while protection of human rights is a guiding principle of the Declaration, it does not refer to the human rights of future generations. Instead, it uses the language of 'needs and interests' of future generations. This represents a watering down of the language in the Zero Draft of the Declaration, which had made several references to future generations' rights. For example, one of the guiding principles originally included in the draft

[68] United Nations Secretary-General, *Intergenerational Solidarity and the Needs of Future Generations* (General Assembly 68th sess, 5 August 2013) <https://sustainabledevelopment.un.org/content/documents/2006future.pdf>.

[69] António Guterres, *Our Common Agenda: Report of the Secretary-General* (2021) <www.un.org/en/content/common-agenda-report/assets/pdf/Common_Agenda_Report_English.pdf>; see also António Guterres, *To Think and Act for Future Generations* (Policy Brief No 1, March 2023) <www.un.org/sites/un2.un.org/files/our-common-agenda-policy-brief-future-generations-en.pdf>.

[70] United Nations, 'Summit of the Future: Multilateral Solutions for a Better Tomorrow' (2024) <www.un.org/en/summit-of-the-future>.

[71] UN Summit of the Future, *Pact for the Future, Global Digital Compact and Declaration for Future Generations* (Outcome Documents of the UN Summit of the Future, 22–23 September 2024).

said: 'The rights of future generations to thrive in prosperity and achieve sustainable development must be guaranteed, including by addressing intergenerational transmission of poverty and hunger, and inequality and injustice'.[72] One of the proposed commitments in the draft was that states would 'Guarantee the rights of future generations to the highest standard of health, through universal healthcare, to ensure healthy lives and promote well-being for all throughout the life course'.[73] Instead, the Declaration refers only briefly to human rights, and avoids any acknowledgement of the rights of future generations specifically. It also appears to avoid the language of a 'human right to a clean, healthy and sustainable environment', apparently walking back from the resolution adopted by the General Assembly in 2022.

The Declaration had potential to recognise the rights of future generations and articulate the obligations which states bear with respect to those rights. Even as a non-binding instrument, this would have been a significant step forward for future generations' rights. Unfortunately, the final version of the Declaration passed by states failed to live up to that potential.

By far the most detailed development in soft law on the rights of future generations and intergenerational equity has come from outside the United Nations. The Maastricht Principles on the Human Rights of Future Generations are a set of guiding principles drafted by a network of civil society advocates and academic experts on human rights and intergenerational equity.[74] The Principles are clear in their position that future generations have human rights which entail obligations on current generations, and they articulate the nature of these obligations in some detail. They will be discussed in greater detail in later chapters as a source of useful principles for how human rights law could be developed to better protect future generations' environmental rights.

3.4 ENVIRONMENTAL RIGHTS AND FUTURE GENERATIONS IN DOMESTIC LAW

While the focus of this chapter is on international law, there are a few notable initiatives at the domestic level which have been influential in the

[72] United Nations Summit of the Future, 'Zero Draft of the Declaration on Future Generations' [13] <www.un.org/sites/un2.un.org/files/co-facilitators_zero_draft_of_the_declaration_on_future_generations_26_march_2024_final_.pdf>.

[73] Ibid [27].

[74] 'Maastricht Principles on the Human Rights of Future Generations' 1 <www.ohchr.org/sites/default/files/documents/new-york/events/hr75-future-generations/Maastricht-Principles-on-The-Human-Rights-of-Future-Generations.pdf>.

advancement of both environmental rights and intergenerational equity. One of the earliest cases to deal with intergenerational environmental rights was *Minors Oposa v Factoran*.[75] In that case, the Supreme Court of the Philippines recognised that the child plaintiffs had a right to a 'balanced and healthful ecology'. This right was not explicitly included in the national constitution but was viewed to be an inherent basic right necessary for self-preservation. The Supreme Court also ruled that the right had an intrinsic intergenerational dimension, noting that each generation has a responsibility towards future generations to ensure they can enjoy the same rights. This was the basis upon which the children were granted standing and was upheld as a duty on the state.

Minors Oposa was influential in linking the environment to fundamental human rights, as well as articulating the obligations owed to future generations.[76] The case followed in the wake of the Rio Summit and the recognition of sustainable development as a key principle of international environmental law. It can be viewed as a bridge between that somewhat aspirational concept and enforceable human rights.[77]

Around twenty-five years later, the Supreme Court in Colombia heard the case of *Demanda Generaciones Futuras v Minambiente* (*Future Generations v Ministry for Environment*).[78] The case was brought by twenty-five young people aged between seven and twenty-five. The young plaintiffs used a special communication procedure under the Colombian Constitution, known as *tutela*, which is based on the concept of guardianship.[79] This enabled them to bring a claim not only in their own names but also on behalf of future generations.

They claimed that the government was in breach of its constitutional obligations by failing to prevent logging of the Amazon rainforest. The court agreed, and held that the contribution to global heating that arose from the

75 *Minors Oposa v Secretary of the Department of Environment and Natural Resources* (1994) 33 ILM 173.

76 Ted Allen, 'Note. The Philippine Children's Case: Recognizing Legal Standing for Future Generations' (1993) 6(3) *Georgetown International Environmental Law Review* 713; Sumudu Atapattu, 'Intergenerational Equity and Children's Rights: The Role of Sustainable Development and Justice' in Claire Fenton-Glynn (ed), *Children's Rights and Sustainable Development: Interpreting the UNCRC for Future Generations* (Cambridge University Press, 2019) 167.

77 United Nations Environment Programme (n 65).

78 *Demanda Generaciones Futuras v Minambiente* [2018] (11001-22-03-000-2018-00319-01, 5 April 2018).

79 Translated by Marcia W Coward, Peter B Heller and Anna I Vellve Torras, 'Colombia's Constitution of 1991 with Amendments through 2005' 130, art 86.

deforestation amounted to a breach of the right to a healthy environment in the Colombian constitution, which the court interpreted to extend to non-human species and ecosystems.[80] This led to one of the two significant findings of the case: the recognition that the Amazon itself has legal rights and was entitled to protection. The other important conclusion was the recognition of the environmental rights of future generations. The court's judgment stated: "The protection of fundamental rights not only involves the individual but implicates the 'other'. The neighbour is otherness; its essence, the other people that inhabit the planet, also includes other animal and plant species. But in addition, this includes the unborn, who also deserve to enjoy the same environmental conditions that we have."[81]

Colombia is one of the more than 100 countries which now recognise environmental rights in some form in their constitutions.[82] Many of these refer to future generations specifically, or to intergenerational equity, sustainable development or some other principle which implies obligations towards future generations.[83] *Demanda Generaciones Futuras* shows the importance of judicial interpretation that recognises intergenerational equity and solidarity with future generations.

Outside of judicial decisions, some of the most significant advancements in protecting the interests of future generations have been through legislative and executive processes. The most well-known of these is in Wales. The *Well-being of Future Generations (Wales) Act 2015* creates a framework of goals relating to sustainability, health, equality, culture and language. Any new law or government decision must consider how it will impact on these domains and seek ways to advance the well-being goals. The law also creates the office of the Commissioner for Future Generations, who is empowered to review how well a public body is performing its obligations under the legislation and can make recommendations on how it should improve its performance. Several other countries have established mechanisms or institutions to advance the interests of future generations in some form, though their functions are mostly supervisory or advisory. The Welsh model is typically put forward as the gold

80 *Demanda Generaciones Futuras v Minambiente* (n 78) [34].

81 Ibid [5.2] per Judge Luis Armando Tolosa Villabona.

82 Slobodian (n 55) 572; James May and Erin Daly, *Global Environmental Constitutionalism* (Cambridge University Press, 2015); Erin Daly and James May (eds), *Implementing Environmental Constitutionalism: Current Global Challenges* (Cambridge University Press, 2018).

83 Slobodian (n 55) 572.

standard for other countries to aspire to when implementing some form of institution for the benefit of future generations.[84]

While these are extremely promising initiatives for bringing future generations' interests to the forefront of governmental decision-making, they are not human-rights-based mechanisms. To truly safeguard the human rights of future generations, these need to be recognised as such, so that they can exist on the same plane as the rights of current generations. To the extent that rights are 'trumps', as Dworkin put it,[85] future generations' interests will always remain vulnerable to limitations and trade-offs if they are not recognised as human rights.

3.5 CONCLUSION

There has been considerable development in environmental rights over the past few decades. We now have widespread recognition of the environmental dimensions or preconditions of other human rights. The right to a healthy environment is recognised in the domestic laws of many states and has been endorsed by a large majority of UN member states in the General Assembly. It is not yet binding at the international level, however. Similarly, intergenerational equity has been recognised in many international instruments but is under-developed as a binding legal principle. The link between environmental rights and future generations has best been articulated by the Inter-American Court of Human Rights, but other courts and tribunals have been slow to confirm this connection. Chapter 4 will explain in more detail why international human rights law has been unable to do more to incorporate intergenerational equity into the enforcement of human rights.

[84] Jonathan Boston, *Governing for the Future: Designing Democratic Institutions for a Better Tomorrow*, 1st ed (Emerald Publishing, 2016) ('*Governing for the Future*'); Iñigo González-Ricoy, 'Constitutionalizing Intergenerational Provisions' in Iñigo González-Ricoy and Axel Gosseries (eds), *Institutions for Future Generations* (Oxford University Press, 2016) 170; Iñigo González-Ricoy and Axel Gosseries, 'Designing Institutions for Future Generations' in Iñigo González-Ricoy and Axel Gosseries (eds), *Institutions For Future Generations* (Oxford University Press, 2016) 3; Jan Linehan and Peter Lawrence, *Giving Future Generations a Voice: Normative Frameworks, Institutions and Practice* (Edward Elgar, 2021).

[85] Ronald Dworkin, *Taking Rights Seriously* (Harvard University Press, 1978).

4

The Problem

Why International Human Rights Law Does Not Protect Future Generations

4.1 INTRODUCTION

The work in this book is premised on the claim, explained in Chapter 2, that international human rights law (IHRL) should protect the environmental rights of future generations. Chapter 3 gave an overview of the content of IHRL and other related frameworks to illustrate some of the gaps in the current law. This chapter complements that work by explaining a number of other barriers that exist in IHRL, fleshing out the reasons why the interests of future generations are not adequately protected. By analysing the various reasons why IHRL struggles to protect the rights of future generations, this chapter will build a foundation for Chapter 5, which seeks to articulate an approach for overcoming, or at least mitigating, these challenges.

As will be shown, there are many factors which contribute to this failure of protection. First and foremost, existing international human rights law does not recognise future generations as holders of rights, largely because the law only recognises states' obligations with respect to people under their jurisdiction or control. While notions of jurisdiction are evolving and may one day encompass obligations across a broader geographical and temporal scope, to date they have not been interpreted to establish duties owed to unborn generations.

Further, there are no specific pathways for future generations' claims to be brought before international human rights bodies, for example, through guardians or other representatives, as exists in some domestic jurisdictions. It is possible for children and young people to pursue human rights complaints, and these claims have great potential for addressing many pressing environmental challenges, but this chapter will argue that children are not an adequate or appropriate proxy for future generations.

The protection of future generations' rights is also limited by the conventional approach to responsibility within human rights law, which adopts a backward-looking liability model where actionable claims are typically limited to actual harms that have already occurred, with some limited extension to cover threats of imminent harms. States do have obligations to prevent interferences with human rights, but again these are usually applied in circumstances where the threat to rights is more immediate and direct. Claims based on more remote and generalised threats of future harm would normally not be considered justiciable, and when they do come before the courts it can be difficult to establish that a state has breached its preventive duties.

A further issue which makes it difficult to uphold the rights of future generations within existing law relates to the inevitable trade-offs which occur between the rights of future and current generations. Given that many environmentally harmful projects are advanced for the stated purpose of promoting economic development, impairment of future generations' interests can often be offset against benefits to current generations. This sort of balancing is common in human rights jurisprudence, where limits to rights can be justified where they achieve other positive human rights objectives, but it is susceptible to the same 'short-termism' that has been identified in law and policy-making more generally.[1] It is difficult to strike a fair balance between current and future interests without a principled approach which ensures future generations' rights are given proper weight. Along with the other reasons noted earlier, this chapter argues that the lack of a meaningful concept of intergenerational justice is a major cause of international human rights law's failure to assure the environmental rights of future generations. Chapter 5 will present a suggestion for what an approach based on intergenerational justice might look like, but first this chapter will unpack the existing barriers in more detail.

4.2 RIGHTS-HOLDERS AND DUTY-BEARERS: FUTURE GENERATIONS ARE NOT HOLDERS OF RIGHTS AND STATES DO NOT OWE THEM DUTIES

A principal barrier to the effective protection of future generations' rights within IHRL is that the law does not expressly recognise them as holders of

[1] Michael K MacKenzie, *Institutional Design and Sources of Short-Termism* (Oxford University Press, 2016); Jonathan Boston, *Governing for the Future: Designing Democratic Institutions for a Better Tomorrow* (Emerald, 2016).

human rights. Human rights are universally enjoyed by all people, but legal protections are based on recognised relationships of rights-holders and corresponding duty-bearers. Put another way, within IHRL rights are enjoyed by people to whom recognised duties are owed. Future generations are not included among the people to whom states owe obligations. Under existing IHRL, governments' obligations are generally owed to persons under their jurisdiction or control.[2] This is usually understood to mean 'effective control' and has been interpreted as placing territorial limits on states' duties.[3] It also implicitly imposes temporal limits which would exclude future generations, since states do not exercise 'control' over persons who do not yet exist.

There has been considerable work in human rights scholarship to encourage a broader understanding of what it is to be under the jurisdiction of a state and thereby expand the situations in which states owe obligations to respect, protect and fulfil human rights. A particular focus of this work has been establishing a basis for extraterritorial human rights obligations, recognising the great potential for states' actions to have negative human rights impacts beyond their borders.[4]

In its General Comment 31, the Human Rights Committee (the body which oversees compliance with the International Covenant on Civil and Political Rights (ICCPR)), explored the scope of states' obligations. The Committee considered that a person would be 'subject to the jurisdiction' of a state if they are within that state's power or control, even if they are outside its

[2] See e.g. the language in the preamble to the *Universal Declaration of Human Rights* (UDHR), which refers to states' duties to protect rights 'among the peoples of Member States themselves and among the peoples of territories under their jurisdiction'. UDHR, General Assembly Res 217A (III) (10 December 1948). The *International Covenant on Civil and Political Rights* (ICCPR) similarly provides in article 2: 'Each State Party to the present Covenant undertakes to respect and to ensure to all individuals within its territory and subject to its jurisdiction the rights recognized in the present Covenant'. ICCPR, 999 UNTS 171, opened for signature 16 December 1966, entered into force 23 March 1976.

[3] Human Rights Committee, General Comment No 31, 'The nature of general legal obligations imposed on states parties to the Covenant' (80th session, 2004), UN Doc CCPR/C/21/Rev.1/Add.13. See also Human Rights Committee, General Comment No 36, 'Article 6: the right to life', which extends the state's obligations to 'all persons subject to its jurisdiction, that is, all persons over whose enjoyment of the right to life it exercises power or effective control'. UN Doc CCPR/C/GC/36 (30 October 2018) [63]; *Al-Skeini and Others v UK*, Application No 55721/07 (Eur Court HR, 7 July 2011).

[4] See e.g. Marko Milanovic, *Extraterritorial Application of Human Rights Treaties* (Oxford University Press, 2011); Nehal Bhuta (ed), *The Frontiers of Human Rights: Extraterritorial Application and Its Challenges* (Oxford University Press, 2016); Karen da Costa, *The Extraterritorial Application of Selected Human Rights Treaties* (Brill, 2013); Tillmann Altwicker, 'Transnationalizing Rights: International Human Rights Law in Cross-Border Contexts' (2018) 29(2) *European Journal of International Law* 581.

territory.[5] This idea of jurisdiction based on control has been applied to extend obligations in cases of occupied territories, offshore refugee detention and extradition cases.[6] It is most easily established where there is physical control over territory or persons or at least physical presence of the duty-bearing state or its agents. In this way, when a state physically exercises its jurisdiction beyond its borders, it bears human rights duties as if it were acting within its own territory.

More recently, some courts and tribunals have adopted more expansive concepts of jurisdiction and control which give rise to extraterritorial obligations even in the absence of physical control of territory. For instance, a variation of the effects doctrine has been used to recognise states' duties to protect the rights of persons outside their borders. The effects doctrine suggests that, where the exercise of a state's jurisdiction will produce negative effects, those actions ought to be subject to human rights obligations.[7] This has been argued as a way of overcoming territorial limits on states' obligations, for example, in the context of transboundary environmental harm which impacts human rights.[8]

In a recent advisory opinion on environment and human rights, the Inter-American Court of Human Rights stated that the exercise of jurisdiction, and consequently the duty to respect and ensure human rights, arises when the state exercises effective control over the activities that caused the

5 Human Rights Committee, *General Comment No. 31: The Nature of the General Legal Obligation Imposed on States Parties to the Covenant* (29 March 2004) 10; Robert McCorquodale and Penelope Simons, 'Responsibility beyond Borders: State Responsibility for Extraterritorial Violations by Corporations of International Human Rights Law' (2007) 70(4) *Modern Law Review* 598; John Knox, 'Climate Change and Human Rights Law' (2009) 50 *Virginia Journal of International Law* 202.

6 E.g. *Loizidou v Turkey*, Application No 15318/89 (Eur Court HR, 23 March 1995), concerning the occupation by Turkish troops of an area of Northern Cyprus; or *Ilaşcu and Others v Moldova and Russia*, Application No 48787/99 (Eur Court HR, 8 July 2004), *where jurisdiction was based on Russia's effective control over the areas based on political, military and economic support. Compare these to Banković v Belgium, where human rights obligations were said not to exist because NATO's bombing of territory of the Former Yugoslavian Republic did not amount to effective control: Banković and Others v Belgium and 16 Other Contracting States*, Application No 52207/99 (Eur Court HR, 12 December 2001). See also *Advisory Opinion on the Legal Consequences of the Construction of a Wall in the Occupied Palestinian Territories* (International Court of Justice, 9 July 2004).

7 See e.g. *Chiragov and Others v Armenia*, Application No 13216/05 (Eur Court HR 16 June 2015) [167]; Altwicker (n 4) 581–606.

8 Knox, 'Climate Change and Human Rights Law' (n 5) 203–204; Alan Boyle, 'Human Rights and the Environment: Where Next?' (2012) 23(3) *European Journal of International Law* 613, 638–640; Human Rights Committee, 'General Comment No. 31: The Nature of the General Legal Obligation Imposed on States Parties to the Covenant' (n 5).

environmental damage, even where that damage occurs in another state's territory.[9] It is not necessary to show effective control over the consequences or the territory where they occur, just over the original act.[10] This makes it potentially more useful for future generations' claims than earlier interpretations of the effects doctrine, which have focused on control over persons or territory.[11]

The reasoning of the Inter-American Court was endorsed by the Committee on the Rights of the Child in its decision in *Sacchi et al v Argentina et al*. That complaint was brought by sixteen young people, including well-known climate activist Greta Thunberg. The petitioners argued that the respondent states (Argentina, Brazil, France, Germany and Turkey) had violated children's rights to life, health and culture under the *Convention on the Rights of the Child* (CRC) through their failure to reduce greenhouse gas emissions.[12] Because the petitioners were mostly not nationals of the respondent states, an issue arose about whether the states owed them any legal obligations that the Committee could enforce. While the Committee ultimately ruled that the complaint was inadmissible owing to the petitioners' failure to exhaust their options for redress under domestic law, it did accept that states could owe obligations to people beyond their borders, such that the respondent states could in theory be responsible for climate harms suffered by the young petitioners.

The Committee articulated three elements for establishing extraterritorial obligations under the CRC.[13] The first requirement is that the state has exercised effective control over activities (in this case the generation of greenhouse gas emissions within its territory). Second, a causal link must exist between the activities and the negative human rights impact (the many harmful effects of climate change). And third, the negative human rights

9 Inter-American Court of Human Rights, *Environment and Human Rights: Advisory Opinion OC-23/17* (15 November, 2017) (Official summary issued by the Inter-American Court, 15 November 2017), <www.corteidh.or.cr/docs/opiniones/resumen_seriea_23_eng.pdf>.

10 Christopher Campbell-Duruflé and Sumudu Anopama Atapattu, 'The Inter-American Court's Environment and Human Rights Advisory Opinion: Implications for International Climate Law' (2018) 8(3–4) *Climate Law* 321–337, 334.

11 *Banković and Others v Belgium and Others*, Application No 52207/99 (Eur Court HR, 12 December 2001); *Al Skeini and Others v the United Kingdom*, Application No 55721/07 (Eur Court HR, 7 July 2011); *Andreou v Turkey*, Application No 45653/99 (Eur Court HR, 3 June 2008); *Advisory Opinion on the Legal Consequences of the Construction of a Wall in the Occupied Palestinian Territories* (International Court of Justice, 9 July 2004).

12 *Sacchi et al v Argentina, Brazil, France, Germany and Turkey* (Communication to the Committee on the Rights of the Child, 23 December 2019).

13 *Sacchi et al v Germany* [2021] Committee on the Rights of the Child, Communication No 107/2019, para 9.7.

impact must have been foreseeable (in this case because of the established climate science).

In relation to the third element of foreseeability, the petition had argued that the sixteen young claimants were the foreseeable victims of climate change caused by the respondent states' failure to adequately reduce greenhouse gas emissions. The foreseeability of climate change impacts was enough, they argued, to bring them within the respondents' jurisdiction and consequently to establish both obligations owed to them under the Convention and standing to bring a complaint.

Foreseeability has been an important component of other attempts to redefine jurisdiction and extend human rights obligations extraterritorially. In articulating the nature of states' duties to prevent transboundary impacts on the right to life, the United Nations Human Rights Committee has explained that states are obliged to prevent 'direct and reasonably foreseeable impacts', even when they occur beyond their borders.[14] With specific reference to climate change, a joint statement by five United Nations human rights committees (including the Committee on the Rights of the Child) states that:

> State parties have obligations, including extra-territorial obligations, to respect, protect and fulfil all human rights of all peoples. Failure to take measures to prevent foreseeable human rights harm caused by climate change, or to regulate activities contributing to such harm, could constitute a violation of States' human rights obligations.[15]

The Committee on the Rights of the Child also addressed foreseeable environmental harm in its General Comment 26. The General Comment was the product of an unprecedented level of consultation, with a special focus on getting input from children. This process occurred concurrently with the complaint in *Sacchi*, and in it we can see a consistent approach. However, the Committee arguably goes one step further in terms of states' duties to prevent foreseeable human rights impacts, hinting at duties to future generations. While it does not go so far as to recognise the rights of future generations, it says that: 'Beyond their immediate obligations under the Convention with regard to the environment, States bear the responsibility for foreseeable environment-related threats arising as a result of their acts or omissions now, the full implications of which may not manifest for years or

[14] Human Rights Committee, *General Comment No.* 36 (n 5) para 22.

[15] Office of the High Commissioner for Human Rights, 'Five UN Human Rights Treaty Bodies Issue a Joint Statement on Human Rights and Climate Change', (2019) para 1 <www.ohchr.org/en/NewsEvents/Pages/DisplayNews.aspx?NewsID=24998&LangID=E>.

even decades'.[16] In a section headed 'Intergenerational Equity and Future Generations' the Committee says that it 'recognizes the principle of intergenerational equity and the interests of future generations, to which the children consulted overwhelmingly referred. While the rights of children who are present on Earth require immediate urgent attention, the children constantly arriving are also entitled to the realization of their human rights to the maximum extent'.[17] This is perhaps the closest that a UN human rights body has come to recognising the rights of future generations, and the notion of 'the children constantly arriving' is a novel and useful way of indirectly including future generations in the cohort of children protected by the Convention.

In conjunction with the approach taken in *Sacchi*, the Committee on the Rights of the Child has potentially created an approach that could extend states' obligations not only extraterritorially, but also into the future. Establishing jurisdiction based on foreseeability and effective control could lead to legally binding obligations, provided that future impacts could be considered reasonably foreseeable and a sufficient causal link could be established between the state's actions and the negative effects for future generations. However, while the Committee is an authoritative body on matters pertaining to the interpretation of the CRC, this approach has not yet been applied successfully in a specific complaint and may prove difficult to enforce. This is related to another problem facing future generations: there is no clear pathway for complaints to be brought on their behalf.

4.3 LACK OF STANDING: FUTURE GENERATIONS CANNOT BRING A CLAIM WITHIN IHRL

Section 4.2 explained that future generations are not recognised as holders of rights within international human rights law and that, consequently, it is difficult to establish that states owe obligations to respect and protect their rights. Even where states' duties could be interpreted to include some obligation with respect to future generations' human rights, enforcing those duties remains problematic. To bring a claim under international human rights law a victim must exist who can point to actual or imminent harm. The challenges for establishing victimhood, and therefore admissibility of future generations' environmental claims, relate to both the temporal remoteness of the expected harm and the inability to pinpoint identifiable victims.

[16] Committee on the Rights of the Child, 'General Comment No. 26 on Children's Rights and the Environment with a Special Focus on Climate Change' [11].

[17] Ibid.

4.3.1 *Approaches to Standing in International and Regional Human Rights Regimes*

International and regional human rights regimes establish rules for admissibility of individual complaints. Generally speaking, complaints are admissible when they come from an individual or group of individuals who are victims of a violation committed by a state which owes them duties. In addition to this are other requirements, such as the common conditions that applicants have exhausted their domestic remedies and that complaints are in writing and not anonymous. Establishing 'victim status' is a common threshold issue for standing and admissibility of human rights claims. It is usually established by showing that a person has already been affected by an act or omission of the state, or that such an effect is imminent.[18]

This was explained by the United Nations Human Rights Committee in *EW et al v The Netherlands* when it dismissed a complaint relating to nuclear weapons in 1993, on the basis that the applicants had not been personally impacted by any harm flowing from the state action.[19] The complainants (6,588 Dutch citizens) had brought the claim after the Netherlands government agreed to the deployment of cruise missiles fitted with nuclear warheads on Dutch territory. In dismissing their complaint, the Committee explained that: 'For a person to claim to be a victim of a violation of a right protected by the Covenant, he or she must show either that an act or an omission of a State party has already adversely affected his or her enjoyment of such right, or that such an effect is imminent, for example, on the basis of existing law and/or judicial or administrative decision or practice'.[20] The Committee articulated the question of standing in these terms: 'The issue in this case is whether the preparation for the deployment or the actual deployment of nuclear weapons presented the authors with an existing or imminent violation of their right to life, specific to each of them'.[21]

18 William J Aceves, 'Actio Popularis – The Class Action in International Law' [2003] (1) *The University of Chicago Legal Forum* 353.

19 *EW et al v The Netherlands* [1993] UN Doc CCPR/C/47/D/429/1990 (1993) Human Rights Committee Communication No 429/1990; *Beydon v France* (2005) Vol II UN Doc A/61/40 Human Rights Committee Communication 1400/2005, 642. The Human Rights Committee assesses standing as part of the admissibility rules set out in the *First Optional Protocol to the International Covenant on Civil and Political Rights* 1966 arts 2, 3, and 5(2). This is expanded on in the Human Rights Committee, 'Rules of Procedure of the Human Rights Committee' 90 <http://hrlibrary.umn.edu/hrcommittee/HRC-RULE.htm>.

20 *EW et al v The Netherlands* (n 19) 6.4.

21 Ibid.

A similar outcome occurred in 2006 in *Aalbersberg v Netherlands*, another complaint relating to the potential impact of nuclear weapons. There, the Committee applied the same test and concluded that, although the authors were concerned about the potential impact of the Netherlands' policy with respect to nuclear weapons, they could not point to a sufficient impact to establish they were 'victims', and their complaint was ruled to be inadmissible.

The complaint in *Brun v France* related to France's authorisation of open-field trials of genetically modified (GM) crops. The applicant had been convicted for participating in protests against the trials, specifically by destroying a test crop, but argued that the potential risks of open-field GM trials ought to have been taken into account in his proceedings. He argued that his treatment constituted a breach of his right to participate in public affairs under article 25 of the ICCPR, and that by allowing the GM trials France had breached his right to life (article 6) and to private and family life (article 17). The Committee followed the same approach as in previous cases and found that there was insufficient evidence that the applicant had been specifically impacted by France's position on GM crops, or that such an impact was imminent. While he had been convicted for his protest activity, there was no evidence that he had been prevented from participating in public affairs through other channels.[22]

The Committee on Economic, Social and Cultural Rights adopts a similar approach to the Human Rights Committee, whereby standing is limited to 'real or potential victims', and claims cannot be presented *in abstracto*. This was explained in the decision in *SC and GP v Italy*.[23] There, the Committee emphasised that it is not able to consider the general compatibility of a state's actions with the ICESCR, but can only admit a complaint if there is, at least prima facie, evidence that 'the authors might be actual or potential victims of the violation of a right enshrined in the Covenant'.[24]

These decisions illustrate an important distinction when it comes to future generations' environmental rights: claims cannot be brought in abstract or general terms. This was confirmed in *Andersen v Denmark*, where the Human Rights Committee followed *Brun v France* and reiterated that a person cannot

[22] *Brun v France* [2006] Human Rights Committee Communication No 1453/2006, 6.3–6.4.

[23] *SC and GP v Italy* [2019] Committee on Economic, Social and Cultural Rights, Communication No 22/2017.

[24] Ibid 6.15.

claim victimhood in 'theoretical terms' or on behalf of a collective interest – they must show that there were specific consequences which personally affected them.[25] The complaint in *Andersen* related to Islamophobic remarks made in the Danish Parliament, which the applicant claimed amounted to a breach of her rights. The Committee found there had been no specific consequences for the applicant and no evidence that she would be personally affected, and ruled that the complaint was inadmissible.

The complaints process before the Human Rights Committee is therefore not open to actio popularis or public interest claims.[26] In *EW et al v The Netherlands*, the Committee put it this way: '[T]he Committee notes that the procedure laid down in the Optional Protocol was not designed for conducting public debate over matters of public policy, such as support for disarmament and issues concerning nuclear and other weapons of mass destruction'.[27] General environmental concerns can therefore not come before the Committee unless they reveal a specific impact on the individual applicants.

A complaint will also be inadmissible if it does not reveal that a harm has either actually occurred or is imminent. If actual harm has occurred, this requirement is seldom controversial and victim status is generally easy to establish. It could be more difficult to prove standing where the complaint relates to a risk of harm materialising in the future, in which case it must be established that the harm is imminent. It is important to note here that imminence may need to be considered both as part of the test for admissibility and again when assessing the merits of the claim – that is, whether a violation has been committed.

The Human Rights Committee decision in *Teitiota v New Zealand* illustrates these different ways of approaching imminence. Mr Teitiota was a national of Kiribati who was seeking refugee protection from New Zealand on the grounds that climate change would cause him harm should he be sent back to his home country. In assessing whether his claim was admissible, the Committee explained that an imminent harm must be more than a theoretical possibility. It concluded that there was prima facie evidence that climate change was impacting negatively upon Kiribati and its people, and that there was therefore an imminent risk of harm flowing from the decision to deport him. However, when it came to assessing the substantive question of whether

[25] *Andersen v Denmark* [2010] Human Rights Committee Communication No 1868/2009; see also *Brun v France* (n 22).

[26] *Andersen v Denmark* (n 25) 6.3; *Brun v France* (n 22) 6.3.

[27] *EW et al v The Netherlands* (n 19) 6.2.

there had been a breach of Mr Teitiota's rights, the Committee held there was insufficient evidence of likely harm to trigger New Zealand's duty of protection. There was not a real and foreseeable risk of harm at the time of the decision. In particular, the Committee pointed to the fact that the timeframe for many of the foreseeable impacts of climate change still allowed time for protective measures to be put in place. On that basis, the Committee ruled that the immigration tribunal's decision not to grant refugee protection was not a breach of Mr Teitiota's rights under the ICCPR.

In *Sacchi*, the Committee on the Rights of the Child considered the admissibility of the children's claim. While some of the allegations related to future climate harms, the Committee relied on the fact that the applicants had 'prima facie established that they have personally experienced a real and significant harm in order to justify their victim status'.[28] The fact that they had already experienced climate harms was sufficient for the claim to be admissible, although it was ultimately rejected on the basis that they had failed to exhaust their local remedies. Had the case been based purely on future harms, it is unclear whether the Committee would have been so willing to find that victim status was made out.

Regional human rights bodies use variations on this method of determining standing and admissibility, but some recent developments could point the way to a more flexible approach. Historically, the European Court of Human Rights (ECtHR) has adopted an approach to admissibility which is very similar to that of the United Nations human rights committees. Under article 34 of the European Convention, the ECtHR can hear applications from any 'person, non-governmental organisation or group of individuals' claiming to be a victim of a human rights violation. For an applicant to have standing based on victim status, they must show that they have been personally affected. Like the UN Human Rights Committee, the European Court does not accept actio popularis. In environmental cases, a general concern for the environment will not be sufficient; some personal impact must be demonstrated.[29] Victims can be directly affected, indirectly affected or potentially affected, but in all cases there must be a clear connection between the alleged harm and the applicants themselves.[30] While it is possible for this to be based on the risk of future harm, the ECtHR will usually hold the case to be inadmissible unless the applicant faces a 'serious,

[28] *Sacchi et al v Brazil* (n 13) para 10.14.

[29] *Di Sarno et al v Italy* [2012] Eur Court HR, Application No 30765/08.

[30] *Mansur Yalçın and Others v Turkey* [2014] Eur Court HR, Application No 21163/11, [40]; *Verein KlimaSeniorinnen Schweiz v Switzerland* [2024] Eur Court HR, Application No 53600/20 [463].

specific and imminent danger'.[31] The harm must be 'not a remote risk'[32] and 'not speculative'.[33]

However, in a recent decision the ECtHR has significantly broadened its approach to standing for cases relating to climate change. In *Verein KlimaSeniorinnen Schweiz v Switzerland*, the court was considering a claim by the not-for-profit organisation Verein KlimaSeniorinnen Schweiz (Senior Women for Climate Protection Switzerland, often referred to as the 'Swiss Grannies') and four of its individual members. Together, they argued that Switzerland had failed to take adequate steps to reduce emissions, thus contributing to heat waves and other environmental impacts which violated their rights to life and to private and family life under the ECHR.[34]

The court was clear that the general exclusion of actio popularis had to be respected, but at the same time recognised that there are special features of climate change which need to be taken into account. Interpreting its previous jurisprudence, the court rejected the claims of the four individual applicants on the basis that they could not show that they had been personally affected. It articulated a two-step test for individual victim status in climate change cases. Firstly, the applicant must be subject to high-intensity exposure to adverse climate impacts. Secondly, there must be a pressing need to establish individual protection owing to the absence or inadequacy of reasonable measures to reduce harm.[35] It found that these conditions were not met by the individual applicants.

However, the court took a different approach for the non-governmental organisation. It recognised that climate change is a common concern of humankind, with an indefinite number of people who have been and could

[31] *Balmer-Schafroth and Others v Switzerland*, Application No 22110/93 (Eur Court HR, 26 August 1996); See also *Fadeyeva v Russia*, Application No 55723/00 (Eur Court HR, 2005); Ole W Pedersen, 'The European Court of Human Rights and International Environmental Law' in John H Knox and Ramin Pejan (eds), *The Human Right to a Healthy Environment* (Cambridge University Press, 2018) 86; Ole W Pedersen, *European Court of Human Rights and Environmental Rights* (Edward Elgar, 2019); Natalia Kobylarz, 'The European Court of Human Rights: An Underrated Forum for Environmental Litigation' in Helle Tegner Anker and Birgitte Egelund Olsen (eds), *Sustainable Management of Natural Resources: Legal Instruments and Approaches* (Intersentia, 2018).

[32] *Budayeva and Others v Russia* [2008] Eur Court HR, Application Nos 15349/02, 21166/02, 20058/02, 11673/02 and 15343/02.

[33] *Soering v the United Kingdom* [1989] Eur Court HR 14038/88.

[34] *Verein KlimaSeniorinnen Schweiz v Switzerland* (n 30).

[35] Ibid [487]; Maria Antonia Tigre and Maxim Bönnemann, 'The Transformation of European Climate Change Litigation: Introduction to the Blog Symposium', *Climate Law Blog* (9 April 2024) <https://blogs.law.columbia.edu/climatechange/2024/04/09/the-transformation-of-european-climate-change-litigation-introduction-to-the-blog-symposium>.

be directly affected. Furthermore, climate change raises significant issues of justice, and the people who are most likely to be affected are often at a 'distinct representational disadvantage'.[36] The court therefore considered it important that organisations can take collective action in the name of affected groups. The court said:

> Because of these fundamental differences, it would be neither adequate nor appropriate to follow an approach consisting in directly transposing the existing environmental case-law to the context of climate change. The Court considers it appropriate to adopt an approach which both acknowledges and takes into account the particularities of climate change and is tailored to addressing its specific characteristics. In the present case, therefore, while drawing some inspiration from the principles set out in the Court's existing case-law, the Court will seek to develop a more appropriate and tailored approach as regards the various Convention issues which may arise in the context of climate change.[37]

At the same time, the court recognised the need to ensure 'that the criteria for victim status do not slip into de facto admission of actio popularis'.[38] For organisations to have standing to bring a claim on behalf of affected individuals or groups, they must be lawfully established and genuinely qualified to represent the people affected. The court noted that it is not essential for the organisation itself to have victim status in order to have standing, but it must show that it represents people who are 'subject to specific threats or adverse effects of climate change on their life, health, well-being and quality of life as protected under the Convention'.[39] For KlimaSeniorinnen, the affected people were older women, who are known to be especially vulnerable to the effects of heat waves and other climate impacts. In this way, the ECtHR has arguably lowered the bar for standing for claims relating to climate change by shifting away from the language of serious, specific and imminent danger. However, it remains to be seen how it would approach environmental rights claims outside the specific facts of the KlimaSeniorinnen case.

Within the Inter-American regional human rights system, individuals may bring a complaint to the Inter-American Commission on Human Rights, though not directly to the Inter-American Court, which can only hear complaints from states or matters referred by the Commission.[40] To satisfy the

36 *Verein KlimaSeniorinnen Schweiz v Switzerland* (n 30) [484], [489].
37 Ibid [422].
38 Ibid [484].
39 Ibid [524].
40 *American Convention on Human Rights* 1969 arts 41(f) and 61(1).

requirements of victim status, a petitioner must have suffered an injury caused by an act which is imputable to the state.[41] The procedural requirements for an individual to commence a claim are relatively flexible but, as the Commission has explained:

> The liberal standing requirement of the inter-American system should not be interpreted . . . to mean that a case can be presented before the Commission *in abstracto*. An individual cannot instigate an *actio popularis* and present a complaint against a law without establishing some active legitimation justifying his standing before the Commission. The applicant must claim to be a victim of a violation of the Convention . . . It is not sufficient for an applicant to claim that the mere existence of a law violates her rights under the American Convention, it is necessary that the law have been applied to her detriment.[42]

The Inter-American Court considers a much wider range of issues in its advisory jurisdiction, where it has been asked to provide opinions on the rights of future generations in the past. In its 2017 Advisory Opinion, the court recognised the right to a healthy environment with both individual and collective dimensions. In its collective dimension, the right is connected to sustainable development and 'constitutes a universal value that is owed to present and future generations'.[43] The court is currently considering another Advisory Opinion, this time on the human rights obligations associated with climate change, and in this context has been asked specifically about obligations to protect future generations and other vulnerable cohorts. In terms of contentious cases, however, some connection to an actual impact is still required to establish standing.

The African human rights system has a different approach to accepting communications alleging breaches of the African Charter on Human and Peoples' Rights. The African Commission on Human and Peoples' Rights may consider actio popularis from individuals and organisations without the need for them to show a personal injury, or even that they 'know or have any relationship with the victim'.[44] As the Commission explained in *Article 19 v*

[41] Aceves (n 18) 384–385.

[42] *Emérita Montoya González v Costa Rica (Admissibility)* (2006) Report No 48/96 OEA/SER.L/V/II.95 Doc 7 Rev at 119 Inter-AmCHR,OEA/SerL/V/II95 Doc 7 Rev at 119 (1997) Case 11.553; Aceves (n 18) 385.

[43] *Environment and Human Rights: State Obligations in Relation to the Environment in the Context of the Protection and Guarantee of the Rights to Life and to Personal Integrity (Advisory Opinion)* [2017] Inter-American Court of Human Rights OC-23/17, 53–55, 59.

[44] *Article 19 v Eritrea* (2007) 73 AHRLR African Commission on Human and Peoples' Rights Communication 275/2003; *Law Society of Zimbabwe and Others v Zimbabwe* [2016] African Commission on Human and Peoples' Rights Communication 321/2006.

Eritrea, 'This is to enable poor victims of human rights violations on the continent to receive assistance from NGOs and individuals far removed from their locality.'[45] This enables a much broader range of matters to come before the Commission and a more meaningful enforcement of environmental rights.[46] However, the members of the Commission have discretion as to which communications they decide to consider, with a majority vote needed before a matter will proceed.[47]

Aside from the African approach, the usual test for standing can be distilled down to two key requirements. The first is a certain degree of specificity of harm facing the applicant – they must be particularly affected as individuals or members of a group of individuals.[48] It is not enough to claim victimhood in 'theoretical terms' or because of a general concern about potential harm that might materialise. To have standing, a victim must show there are specific consequences which personally affect them.[49] The second requirement is that the harm must have actually occurred or be imminent. Actual harm is straightforward enough, but cases relating to harm which has not yet occurred must establish that there is an 'imminent' impact.

The prospect of future environmental harm – which could potentially be large-scale, severe and irreversible – raises questions about both these elements of 'victimhood'. The standard of imminent harm is obviously difficult to make out with truly future harm. But it is also difficult to show particular or personal impact on any given member of future generations.[50] The European Court of Human Rights has potentially moved this approach forward in the recent KlimaSeniorinnen decision. Where a person or group faces severe climate impacts that require state protection, they may now be able to satisfy the test for victim status and can be represented by an organisation to make a human rights claim. But the court was clear that this was an approach based on the specific features of climate change, so it remains at this stage a limited avenue available in Europe.

[45] *Article 19 v Eritrea* (n 44).

[46] Elinor Buys and Bridget Lewis, 'Environmental Protection through European and African Human Rights Frameworks' (2022) 26(6) *International Journal of Human Rights* 949.

[47] *African Charter on Human and Peoples' Rights* 1981 art 55.

[48] *EW et al v The Netherlands* (n 19); *Beydon v France* (n 19); *Brun v France* (n 22); *Aalbersberg v Netherlands* [2006] Human Rights Committee 1440/2005; *Andersen v Denmark* (n 25).

[49] *Andersen v Denmark* (n 25); also *Brun v France* (n 22).

[50] Birgit Peters, 'The European Court of Human Rights and the Environment' in Edgardo Sobenes, Sarah Mead and Benjamin Samson (eds), *The Environment through the Lens of International Courts and Tribunals* (TMC Asser Press, 2022) 189.

International human rights laws are also justiciable before the International Court of Justice (ICJ). The ICJ hears disputes between states.[51] It has no power to hear a complaint from an individual or group of individuals, and human rights-based claims are less common than other areas of international law for this reason. Nonetheless, it is worth considering how a case might come before the ICJ with the objective of enforcing future generations' environmental rights.

There are several ways in which a complaint can come within the jurisdiction of the ICJ. One is where a human rights treaty specifically creates a pathway for disputes to be referred to the ICJ, as is the case with the *Convention on the Prevention and Punishment of the Crime of Genocide*.[52] Other human rights treaties, like the *Convention on the Elimination of All Forms of Racial Discrimination* (CERD), permit referral to the ICJ once the other treaty processes have been exhausted.[53] Jurisdiction can also be established where states agree themselves to refer a dispute to the court for resolution.[54] The third alternative for jurisdiction is a situation where a state has made a general declaration that it consents to jurisdiction with respect to disputes with other states who have made a similar declaration.[55]

As well as establishing that the ICJ has jurisdiction to hear the dispute, an applicant state must also show that it has standing. This is usually on the basis that the state is an injured party, in that it has suffered some negative impact as a result of another state's (or states') breach of obligation. These rules are set out in the Articles on State Responsibility, which are considered a codification of customary international law.[56] In human rights cases this can work in tandem with the doctrine of diplomatic protection, under which a state can bring an action on behalf of an injured national, provided they have exhausted their options for obtaining a remedy at the local level.[57] An alternative basis for standing is through the principle of erga omnes (where an obligation owed is considered important enough that any state may take action to address a breach).[58]

[51] *Statute of the International Court of Justice* 1948 art 34.

[52] *Convention on the Prevention and Punishment of the Crime of Genocide* 1948 art IX; *Statute of the International Court of Justice* (n 51) art 36(1).

[53] *International Convention on the Elimination of All Forms of Racial Discrimination* 1965 art 22 ('CERD').

[54] *Statute of the International Court of Justice* (n 51) art 36(1).

[55] Ibid art 36(2).

[56] *Responsibility of States for Internationally Wrongful Acts* 2001 art 42.

[57] *Draft Articles on Diplomatic Protection* 2006.

[58] *Responsibility of States for Internationally Wrongful Acts* (n 56) art 48; *Barcelona Traction, Light and Power Company, Limited (Belgium v Spain)* [1970] ICJ Rep 3.

It is worth briefly considering the reasons why these rules of standing almost certainly exclude future generations' claims. First, if a state seeks to enforce the rights of its own people through the doctrine of diplomatic protection, it will need to establish both that the persons affected are nationals and that they have suffered an injury as a result of the conduct of another state or states. Article 4 of the Draft Articles on Diplomatic Protection (which also reflect customary international law) specifies that nationality is determined by the laws of the state concerned, usually established either through birth, descent or naturalisation.[59] The possibility of nationality by descent opens some possibility that a state might declare future generations to be 'nationals', but this is normally applied to people already born to parents who are nationals, not to unborn people. Most relevantly, article 4 refers to nationality that a person 'has acquired', indicating it is a retrospective assessment. The doctrine of diplomatic protection has not been applied to include future nationals. Edith Brown Weiss argued that states should be able to bring complaints on behalf of both their current and future nationals, but this has yet to be attempted.[60]

Diplomatic protection is closely linked with the rules of state responsibility, which establish the circumstances in which an injured state may seek to enforce the law against another state. While the Articles on State Responsibility do not specifically address the question of the timing of harm, they do largely frame the rules in terms of an 'injury caused', implying a retroactive effect. The Commentary on the Articles explains that some obligations may be breached before harm occurs if they prohibit threatened or attempted conduct, or if they are preventative in nature.[61] The duty to prevent transboundary harm is an example of such a rule, where a breach might occur based on a failure to take reasonable preventative steps in response to a risk of environmental harm, regardless of whether the harm eventuates. However, the Commentary also clarifies that generally an anticipated breach will not be actionable. Further, while a *breach* of a preventative duty may occur prior to harm occurring, it is the *injury* which gives rise to standing. The injury can be 'material or moral',[62] but some injury must have occurred in order for a state to bring a claim. Overall, the circumstances in which a state could bring an

[59] *Draft Articles on Diplomatic Protection* (n 57) 4.

[60] Edith Brown Weiss, *In Fairness to Future Generations: International Law, Common Patrimony and Intergenerational Equity* (Transnational, 1989) 121.

[61] International Law Commission, *Draft Articles on Responsibility of States for Internationally Wrongful Acts, with Commentaries* (2001) 61–62.

[62] *Responsibility of States for Internationally Wrongful Acts* (n 56) 31.

action for a future injury against its nationals would therefore be extremely limited, absent some specific international obligation to protect or ensure their rights.

The other option for standing in the ICJ is invoking an obligation erga omnes. This depends on the nature of the obligation that is alleged to have been breached. It is available where the obligation is owed to the international community as a whole, or to all the members of a treaty collectively.[63] To date, the ICJ has only recognised a limited number of obligations as being erga omnes in character, including protections against genocide, slavery and racial discrimination.[64] It has also recognised that the right to self-determination gives rise to erga omnes obligations.[65]

To rely on it successfully for the benefit of future generations would require locating an obligation within international law that obliges all states to protect future generations' rights. As discussed in Chapter 3, while intergenerational equity is a principle of international law, and despite the references to that principle and to future generations in several instruments, it has yet to reach the status of a binding international rule which would create obligations erga omnes. Even if intergenerational equity is recognised as a general principle of law or a rule of custom, this would merely enable the ICJ to apply it as a legal rule.[66] Something more is required to elevate it to the kind of obligation which justifies erga omnes status.

Edith Brown Weiss argued that the obligation to protect the global commons, which underpins her theory of intergenerational equity, could be the sort of duty to which erga omnes status applies, but to date this has not been established as a valid legal avenue.[67] In her critique of Brown Weiss's theory of intergenerational equity, Fitzmaurice has said: 'The difficulties, however, with this basis for standing in judicial proceedings, at least before the ICJ, are at the moment impossible to circumvent. At present, international law does not have such a suitable mechanism to accommodate such claims'.[68]

[63] Ibid 48 (known as *erga omnes partes*).

[64] *Barcelona Traction, Light and Power Company, Limited (Belgium v Spain)* (n 58).

[65] *East Timor (Portugal v Australia)* [1995] ICJ Rep 90; *Legal Consequences of the Construction of a Wall in the Occupied Palestinian Territory (Advisory Opinion)* (2004) [2004] ICJ Rep 136 (2004).

[66] *Statute of the International Court of Justice* (n 51) art 38(1).

[67] Brown Weiss, *In Fairness to Future Generations* (n 60) 121; See also Laura Horn, 'Climate Change and the Future Role of the Concept of the Common Concern of Humankind' (2015) 2 *Australian Journal of Environmental Law* 24; *Gabčíkovo-Nagymaros Project (Hungary/Slovakia) (Separate Opinion of Judge Weeramantry)* [1997] ICJ Rep 88.

[68] Malgosia Fitzmaurice, 'Intergenerational Equity Revisited' in Isabelle Buffard et al (eds), *International Law between Universalism and Fragmentation* (Brill Nijhoff, 2008) 228.

As the law currently stands, contentious claims to uphold the human rights of future generations seem very unlikely to be admissible before the ICJ.

4.3.2 *Children's Litigation for Future Generations*

Without an option for standing on behalf of future generations directly, a common approach in environmental rights litigation is for children and young people to pursue a case on their own behalf for harms predicted to occur in their lifetimes, while including arguments relating to future generations as well. We have seen many of these sorts of cases in relation to climate change across a range of national and international jurisdictions. These cases include the *Sacchi* communication, noted earlier, as well as *Duarte Agostinho and Others v Portugal and Others*, a case before the European Court of Human Rights in which a group of young people alleged violations of their rights against thirty-three European countries.[69]

The enforcement of young people's, and especially children's, human rights goes a long way to addressing climate injustice. Considering their young age and the fact many of the anticipated milestones for global warming will occur within their lifetimes, if we can protect their human rights we will have gone a long way to halting the worst effects of climate change. Moreover, more distant generations will inevitably benefit from any successful litigation which results in stronger action from states on climate change. The reality that future generations will be incidental beneficiaries of climate litigation brought by children and young people has been argued to be an adequate substitute for dedicated mechanisms to enforce future generations' rights more directly.[70]

In *Sacchi*, the children argued that states are under a 'duty to ensure intergenerational equity for children and for posterity'. They derived this obligation from a synthesis of the best interests principle and the principle of non-discrimination found in the CRC, along with concepts of stewardship and obligations for posterity found in environmental law and the constitutions of countries like Germany and the United States.[71] In *Duarte Agostinho*, the claims relate principally to the effects of climate change on the young petitioners themselves, but intergenerational equity was presented as a principle to

[69] *Duarte Agostinho and Others v Portugal and 32 Others* [2024] Eur Court HR [GC] (Application No 39371/20).

[70] John H Knox, 'Constructing the Human Right to a Healthy Environment' (2020) 16(1) *Annual Review of Law and Social Science* 79.

[71] *Sacchi et al (Petition)* (n 13) paras 193–195.

which the European Court should have regard in assessing the degree of risk to which they are being exposed.[72]

Another variety of cases to consider future generations are those brought by transgenerational groups, such as Indigenous peoples or other groups who have specific intergenerational rights and responsibilities.[73] Like the children's litigation, these cases generally rely on the standing of the living members of the group, but allow for future generations' interests to be brought before the court because of the cultural relationship between current generations and their ancestors and descendants.

These approaches provide examples of how complaints in human rights courts and tribunals could be used to benefit future generations, even without litigating their rights explicitly. González-Ricoy and Gosseries describe this as making existing institutions more 'future beneficial'.[74] Ultimately, however, the legal opportunities to enforce the rights of future generations will depend on the rules of standing within the relevant legal system.[75] The fact that most of the cases dealing with future environmental harms discussed earlier have been brought by children and young people is indicative of the relatively limited access to courts afforded for future generations specifically. However, there are issues with relying on children and young people to be 'proxies' for future generations.

4.3.3 *The Problem of Proxies*

As the preceding discussion shows, most of the cases which have looked at future generations' rights have been litigated by children and young people. Proxy representation of future generations by children and young people raises a number of issues, however, and it is worth considering whether and to what extent this could be used as a substitute for dedicated enforcement mechanisms.

72 *Duarte Agostinho* (n 18) annex, para 8.

73 Lydia Slobodian, 'Obligations to Transgenerational Groups: A Justification for Sustainable Environmental Policy' (2012) 24 *Georgetown International Environmental Law Review* 387; *Daniel Billy and Others v Australia* [2022] Human Rights Committee Communication No. 3624/2019; *Pabai Pabai & Anor v Commonwealth of Australia* (Federal Court of Australia, VID622/2021).

74 Iñigo González-Ricoy and Axel Gosseries, 'Designing Institutions for Future Generations' in Iñigo González-Ricoy and Axel Gosseries (eds), *Institutions for Future Generations* (Oxford University Press, 2016) 3.

75 Bridget Lewis, 'Protecting Environmental Human Rights for Future Generations' in Walter F Baber and James May (eds), *Environmental Human Rights in the Anthropocene: Concepts, Contexts, and Challenges* (Cambridge University Press, 2023).

Children and young people can act as proxies for future generations in environmental rights litigation in three slightly different ways. The first is explicitly, in the rare instance where standing is available to bring a representative claim specifically on behalf of future generations. These sorts of pathways are available in some countries, such as the Philippines and Colombia, where constitutional rights are recognised to extend to future generations and can be enforced through representative claims (these examples are discussed in more detail in Chapter 7). An explicit claim could be brought with or without the second category, which is an implicit claim on behalf of future generations that is included in the children's own claim. For example, a child could make a claim relating to future harms and draw on principles of intergenerational equity. Success in such a claim would have implicit benefits for future generations, even though the claim was not brought on their behalf explicitly. A third related but conceptually distinct version would be a claim by children for current or imminent harms which does not explicitly involve intergenerational considerations but which nonetheless has instrumental benefits for future generations by leading to a change in state action. The discussion of representative claims here relates to the first two categories, where children either explicitly or implicitly represent the interests of future generations. For various reasons, however, this is not an adequate solution for international human rights law.

Having children and young people act as proxies with the task of representing generations beyond their own raises its own questions about fairness and intergenerational equity. Adults have a moral duty towards children, which is reflected in the overarching requirement to act in the best interests of the child, found in the Convention on the Rights of the Child.[76] In relation to climate change in particular, there is a moral duty which flows from adults' greater contribution to GHG emissions and greater participation in decisions which have got us to this point.[77] To discharge their moral duty, adults ought to take responsibility for enforcing relevant climate obligations, demanding stronger action from states and protecting the human rights of younger generations. Passing part of this responsibility on to children by requiring them to shoulder the burden of enforcing future

[76] Sumudu Atapattu, 'Intergenerational Equity and Children's Rights: The Role of Sustainable Development and Justice' in Claire Fenton-Glynn (ed), *Children's Rights and Sustainable Development: Interpreting the UNCRC for Future Generations* (Cambridge University Press, 2019) 167; Karen E Makuch, 'Environmental Rights of Children' in Michael Faure (ed), *Elgar Encyclopaedia of Environmental Rights* (Edward Elgar Publishing, 2019) 386, 388.

[77] See e.g. Simon Caney, 'Cosmopolitan Justice, Rights and Global Climate Change' (2006) 19(2) *Canadian Journal of Law and Jurisprudence* 255.

generations' rights adds to their already disproportionate (that is, unjust) burden regarding climate change.

Using the language of Edith Brown Weiss's theory of intergenerational equity, we all have a duty to conserve equity of options, quality and access for future generations.[78] If our only means of enforcing this duty is to have children and young people represent future generations' interests as well as their own, then older generations arguably shirk their responsibilities. In a legal setting, including future generations' claims may also make children's and young people's cases more difficult to win, for the reasons outlined earlier.

Such an approach also assumes that today's children have similar or compatible interests to unborn generations, which may not always be the case. As Daly explains, conflating children with future generations can diminish the attention given to children's specific needs and interests. She says that children 'are often seen solely in terms of their future potential rather than present experiences and capacities'.[79] While there may be overlapping interests, we should be conscious of the precise nature of this overlap and not assume that children's interests and those of future generations will necessarily align. While children and young people provide an essential perspective for long-term decision-making, they should not be left the task of representing future generations. As the United Nations Secretary-General said recently as part of his work on Our Common Agenda: 'Young people alive today should not bear the burden of advocating for our descendants, but they are deserving of a separate and dedicated place at the table'.[80]

If their interests do not align then we will need to confront challenging questions of how to balance potentially conflicting rights. It is foreseeable that a legal case brought by children will tend to favour their interests, however, which may leave future generations' rights under-protected. In children's rights law and litigation, the requirement to act in the best interests of the child is a compelling one. However, complying with this in respect of specific child plaintiffs may in many cases lead to conflicts with the best interests of future generations (including future children). Using children as proxies for future generations may therefore set up a normative clash between the best interests principle (which is fundamental to the children's rights framework)

[78] Edith Brown Weiss, *In Fairness to Future Generations* (Transnational, 1989).

[79] Aoife Daly, 'Intergenerational Rights Are Children's Rights: Upholding the Right to a Healthy Environment through the UN Convention on the Rights of the Child' (SSRN Scholarly Paper No 4141475, 20 June 2022) 4 <https://papers.ssrn.com/abstract=4141475>.

[80] António Guterres, *To Think and Act for Future Generations* (No Policy Brief No 1, March 2023) <www.un.org/sites/un2.un.org/files/our-common-agenda-policy-brief-future-generations-en.pdf>.

and intergenerational equity (or other expressions of duty regarding future generations' rights).

Truly ensuring equity of options, quality and access requires that each generation's interests can be given appropriate consideration in their own right. This points to the need for dedicated mechanisms through which future generations' environmental rights can be enforced. Not only is this more likely to produce an outcome which is future-focused, but it also reduces the risk of exacerbating climate injustice by requiring children to take on responsibility for enforcing the rights of future generations. The options for a dedicated means of enforcing future generations' environmental rights will be explored in more detail in Chapter 5.

4.4 ESTABLISHING LEGAL RESPONSIBILITY: THE CHALLENGES OF ACTIONING FUTURE HARMS

Enforcing the rights of future generations within international law is also limited by that system's approach to future or anticipated human rights harms. The issues here are related to those of standing discussed earlier, where future harm was considered in the context of establishing victimhood. Assuming a claim can overcome that hurdle, it will face the further challenge of showing that the respondent state is responsible for a breach of its legal obligations. There are three related but distinguishable aspects to establishing state responsibility in this regard: (1) showing that a state has failed to meet the standards attached to the relevant duty; (2) proving a connection between the state's conduct and the alleged negative impact on human rights; and (3) demonstrating that the state cannot justify the negative impact on the basis of some countervailing legitimate aim.

4.4.1 *Demonstrating a Breach of Duty*

The difficulty of establishing that a respondent state owes a duty to future generations was outlined earlier. Here, the issue relates to showing that such a duty has been breached. As Schmid explains, 'a violation of human rights law is established by showing that there is disconformity between a state's obligations and its actions. This involves asking what measures would have been appropriate and how do the state's actions compare?'[81] To successfully show a

[81] Evelyne Schmid, 'Victim Status before the ECtHR in Cases of Alleged Omissions: The Swiss Climate Case', *EJIL: Talk!* (30 April 2022) <www.ejiltalk.org/victim-status-before-the-ecthr-in-cases-of-alleged-omissions-the-swiss-climate-case>.

breach, the duty needs to be defined with appropriate standards of performance, and evidence must demonstrate that the state has failed to adhere to those standards.

A challenge for litigating the environmental rights of future generations is a lack of sufficiently detailed future-oriented duties and standards. The environmental rights outlined in Chapter 3 are accompanied by duties that can be either negative (that is, requiring states to refrain from doing something) or positive (requiring them to do something). Some of these duties have a preventative, forward-looking dimension, for example, the positive obligation to protect people from threats to life. However, these sorts of duties are usually only triggered by the existence of an imminent threat.

For future generations' claims, the interpretation of imminence is a critical factor which can determine the success or failure of a case. Understanding the role of imminence in existing human rights law is complicated by the fact that it can be relevant at various stages of the litigation. As noted earlier, imminence of harm is commonly a factor in demonstrating victimhood for the purposes of standing. It can also be relevant to whether state obligations of protection are triggered in the context of a foreseeable threat to rights. And, as discussed later, it is further considered when determining whether an alleged failure by the state has caused or is likely to cause a negative human rights violation.

The imposition of a threshold of imminence or immediacy helps to put some limits on states' positive duties which are arguably necessary for practical reasons, but it has the effect of significantly constraining the circumstances in which a breach will be found in relation to foreseeable future harm.[82] Case law in international and regional courts shows that duties of protection are normally only breached where there has been a failure to prevent a specific imminent harm. The test in the European case of *Osman v UK* has been influential in this respect. That case established that the government must take positive steps to safeguard the right to life where it has knowledge of a particular risk to life and where that risk is immediate.[83] Where the danger is more remote, no obligation will apply. The Osman approach has been applied in subsequent cases, both within and beyond Europe.[84]

[82] Justine Bell-James and Briana Collins, 'Human Rights and Climate Change Litigation: Should Temporal Imminence Form Part of Positive Rights Obligations?' (2022) 13(1) *Journal of Human Rights and the Environment* 212.

[83] *Osman v the United Kingdom* [1998] Eur Court HR 23452/94; *Keenan v the United Kingdom* [2001] Eur Court HR 27229/95; Bell-James and Collins (n 82).

[84] *Keenan v the United Kingdom* (n 83); *Case of the Pueblo Bello Massacre v Colombia* (Inter-American Court of Human Rights, 31 January 2006).

There have been some cases where a breach has been found based on a failure to address a foreseeable risk, even though the harm was not necessarily imminent. For example, in *Öneryildiz v Turkey*, the state was found responsible for failing to protect residents from the risk of a methane explosion at a nearby waste dump. Evidence showed that the state had been aware of the risk for some time and failed to take necessary precautions to prevent an explosion. However, the Grand Chamber of the European Court of Human Rights concluded that, even though the risk had been present for over two years, it was still an immediate risk because the explosion could have happened at any time. The state was under an obligation to take urgent steps to prevent the harm from occurring.[85]

Another challenge in articulating duties that might apply for future generations is that the terminology around real, imminent, or immediate threats is not consistently applied. Bell-James and Collins have analysed the way this language has been used in regional human rights courts and found that the terms are often conflated or given different meanings. In particular, they point to the fact that some cases require foresight of a specific and imminent threat,[86] while in others, general knowledge of circumstances of risk were sufficient to trigger duties of protection.[87] They also note differences in language around foreseeability, due diligence and risk, which influence the specific standards of performance that states are expected to adhere to.[88]

These cases show that state duties with respect to future harm will usually only apply where a risk is imminent or immediate, and are much easier to establish where the risk is to an identified individual or group. Generalised threats are much more difficult to enforce. As the threat to rights becomes more 'complex, urgent and largescale'[89] it becomes more difficult to point to a specific duty that has been breached. Bell-James and Collins have critiqued this emphasis on temporal imminence and individualised risk, showing that it can produce unjust outcomes, particularly in environmental rights cases. In environmental rights claims on behalf of future generations, where the risk

[85] *Öneryildiz v Turkey* [2004] Eur Court HR [GC] 27785/10, 101.

[86] See e.g. *Osman v the United Kingdom* (n 83); *Keenan v the United Kingdom* (n 83); *Case of González et al ('Cotton Field') v Mexico* (Inter-American Court of Human Rights, 16 November 2009); *Case of Human Rights Defender et al v Guatemala* (Inter-American Court of Human Rights, 28 August 2014).

[87] See e.g. *Öneryildiz v Turkey* (n 85); *Budayeva and Others v Russia* (n 32); *Case of the Pueblo Bello Massacre v Colombia* (n 84); *Jessica Lenahan (Gonzales) v United States* [2011] Report No. 80/11 (2011) Inter-American Court of Human Rights Case No. 12.626.

[88] Bell-James and Collins (n 82) 221–228.

[89] Schmid (n 81).

of harm may take many years to materialise, it will be difficult to show that a state was under a duty to act in a particular way.

4.4.2 *Connecting a Breach of Duty with Future Harm*

A related issue is that, even where a court accepts that a state was under a duty to take steps in relation to future generations' rights, proving a violation usually requires drawing a connection between the state's conduct and an impact on rights for identifiable people. Again, the case law shows that this link needs to be drawn between the state's actions and a negative impact that either has already occurred or is imminent.[90] Causation of harm is much easier to demonstrate with a close temporal link but becomes more difficult as the anticipated harm becomes more remote in time.

This is illustrated by the decision of the UN Human Rights Committee in *Teitiota v New Zealand*. In that case, the Human Rights Committee considered whether sea-level rise caused by global heating could constitute a violation of the right to life of people in Kiribati.[91] The focus of the case was whether New Zealand's plan to deport Mr Teitiota back to Kiribati breached its duty of protection under the right to life. The Committee explained that, in order for the duty of protection to be triggered, the risk to life 'must be, at least, likely to occur'.[92] While the Committee accepted that rising sea levels threaten the lives of I-Kiritbati, it concluded that no violation had occurred because that threat was not imminent and could be prevented by taking various adaptation actions.

In the more recent case of *Daniel Billy v Australia*, there was a discussion of whether the effects of climate change on communities in the Torres Strait were existing or imminent and whether Australia was in breach of its positive duties with respect to the rights to life, culture and private, family and home life. The majority of the Human Rights Committee found breaches of articles 17 (the right to private, family and home life) and 27 (right to culture) based on impacts that were already occurring to the complainants' lands and waters and Australia's failure to protect against this damage. It found no breach of the right to life, however, because the timeframe of impacts still allowed scope for

90 *EW et al v The Netherlands* (n 91); *Beydon v France* (n 19). The HRC stated that an individual 'must show either that an act or an omission of a State party has already adversely affected his or her enjoyment of such right, or that such effect is imminent' (para 4.3).

91 *International Covenant on Civil and Political Rights*, opened for signature 16 December 1966, 999 UNTS 171 (entered into force 23 March 1976) art 6 ('*ICCPR*').

92 *Ioane Teitiota v New Zealand*, Communication No 2728/2016 (Human Rights Committee, 24 October 2019).

the state to adopt appropriate adaptation measures. While the evidence showed that, without action, the impacts of climate change would certainly worsen in the Torres Strait and that the lands could become uninhabitable within ten to fifteen years, 'The Committee consider[ed] that the time frame of 10 to 15 years, as suggested by the authors, could allow for intervening acts by the State party to take affirmative measures to protect and, where necessary, relocate the alleged victims'.[93]

A similar approach has been taken in regional human rights bodies. In the recent decision of the European Court of Human Rights in Verein KlimaSeniorinnen (discussed earlier), the court explained that for duties of protection to apply in the context of an activity which is, by its very nature, capable of putting an individual's life at risk, there has to be a 'real and imminent' risk to life.[94]

These cases demonstrate that, for rights-based litigation which seeks to combat climate change or other long-term environmental harm, a time delay between state conduct and consequential human rights impacts will be problematic. While there may be compelling evidence that future environmental problems are likely, a causal connection to individualised and imminent impacts will be difficult to demonstrate, especially if a court is of the view that there is still time to avoid a negative impact through other action.[95]

A further problem for showing that a state's actions are causally connected to a negative impact on environmental rights is that, especially for climate change, environmental impacts are often the result of many states' actions, accumulating over time. They can also be influenced by other complex environmental and climate factors. In relation to climate change, it can be extremely difficult to pinpoint a given act, decision or policy as *the cause* of climate harms, so a 'but for' approach to causation is clearly inadequate for such cumulative harms. For environmental harm, the conventional legal concept of a chain of causation has been criticised for ineptly capturing the nature of environmental systems. Fisher explains that

93 *Daniel Billy and Others v Australia* (n 73) 8.8.

94 *Verein KlimaSeniorinnen Schweiz v Switzerland* (n 30) [511].

95 Atapattu (n 76) 174; Meinhard Doelle, 'Climate Change and Human Rights: The Role of the International Human Rights in Motivating States to Take Climate Change Seriously' (2004) 1(2) *Macquarie Journal of International and Comparative Environmental Law* 179, 203–205; John Knox, 'Linking Human Rights and Climate Change at the United Nations' (2009) 33 *Harvard Environmental Law Review* 477, 489; Office of the High Commissioner for Human Rights, *Report of the Office of the United Nations High Commissioner for Human Rights on the Relationship between Climate Change and Human Rights* (15 January 2009) para 70; Ole Pedersen, 'Climate Change and Human Rights: Amicable or Arrested Development?' (2010) 1(2) *Journal of Human Rights and the Environment* 246.

a 'net' of causation is a more appropriate analogy, as it better reflects the complexity of multiple interacting factors that influence the outcome of a decision or action.[96]

The anthropocentric, individualised approach to environmental rights also tends to exclude considerations of the structural, cumulative and complex factors that often cause environmental harm.[97] The recognition of ecocentric rights would likely make these claims easier to substantiate, because harm could be established based on damage to the environment more generally. For example, in the Colombian case of *Demanda Generaciones Futuras v Minambiente*, the existence of an ecocentric right in the Colombian constitution allowed for a more direct argument in terms of the risk to future generations, avoiding the need to show foreseeable or imminent harms to other specific human rights. Perhaps more significantly, the Supreme Court confirmed that the right to a healthy environment protects the inherent value of nature, not just its instrumental value to human interests. Protecting this value entails obligations towards all inhabitants of the planet, both human and non-human. Importantly, the court recognised that this extends to future inhabitants, who are entitled to enjoy the same environmental conditions as current generations.[98] In this way, the court used concepts of intergenerational equity to frame obligations and rights relating to future generations. The Colombian case suggests that environmental rights that are framed in an ecocentric way may be better suited to claims on behalf of future generations because they more easily accommodate notions of intergenerational equity and solidarity, giving rise to obligations to preserve the environment for the benefit of future generations.

With this in mind, the recent resolutions of the UN Human Rights Council and General Assembly (discussed in Chapter 3) may offer some hope that future generations' human rights could be more meaningfully protected at the international level. However, as IHRL currently stands, a causal link to some specific and imminent harm experienced by an identified individual or group would still seem a necessary part of successful litigation.

[96] DE Fisher, *Australian Environmental Law: Norms, Principles & Rules*, 3rd ed (Thomson Reuters, 2014) 644; Applied by Kingham P in *Waratah Coal Pty Ltd v Youth Verdict Ltd and Others (No 6)* [2022] Land Court of Queensland QLC, 21, 1339.

[97] Dinah Shelton, 'Environmental Issues and Human Rights: Multilateral Treaties Adopted between 1991 and 2001', *Joint UNEP-OHCHR Expert Seminar on Human Rights and the Environment* (14–16 January 2002).

[98] *Demanda Generaciones Futuras v Minambiente* [2018] (11001-22-03-000-2018-00319-01, 5 April 2018) 5.2.

4.4.3 *Balancing against Other Human Rights Objectives*

One of the other challenges in establishing a breach of obligation is showing that there is no valid justification for the limitation on future people's rights. Most human rights can be lawfully restricted where it is necessary to do so in order to protect other rights, provided the restriction is proportionate.[99] This balancing of competing human rights is commonly done in an intragenerational context, but can be difficult to do when weighing up the rights of future generations against those of current generations. Currently, international human rights law lacks clear principles and precedents for doing this.

Climate change offers numerous examples of the complexity of this challenge. To safeguard future generations' rights, we know that rapid reductions in greenhouse gas emissions are required. This will require major transformations in the energy, transportation, agricultural and manufacturing sectors. These changes could limit the rights of current generations, like the right to work or the right to the highest attainable standard of living. Emissions reductions achieved through land use changes – for example, switching agricultural production to biofuels or constructing solar or wind farms – potentially raise issues for the rights to property and Indigenous peoples' rights to land and culture.[100]

Currently there is very little guidance for how courts or law-makers should balance these different obligations across timescales. While intergenerational equity has a strong presence in international environmental law, it is almost entirely absent within IHRL. The major human rights treaties (ICCPR and ICESCR) were drafted prior to the emergence of intergenerational equity and sustainable development as influential principles on the international stage. The UN Charter refers to future generations once, in articulating the aspiration of sparing future generations from the scourge of war. Intergenerational equity and future generations are otherwise absent from international human rights law.

In cases relating to future environmental harm, the remoteness of future generations and the uncertainty that surrounds their needs and capacities will inevitably be compared against the much more present and pressing needs of

[99] See e.g. *International Covenant on Civil and Political Rights* 1966 arts 18 and 19 ('*ICCPR*') which permit limitations on the rights to freedom of religion and expression where such restrictions are necessary to protect the rights or freedoms of others.

[100] Bridget Lewis, 'Human Rights' in Gerry Nagtzaam, Katie O'Bryan and Mark Beaufoy (eds), *Legal Pathways to Deep Decarbonisation in Australia* (LexisNexis, 2023); Bridget Lewis, 'Balancing Human Rights in Climate Policies' in Ottavio Quirico and Mouloud Boumghar (eds), *Climate Change and Human Rights* (Routledge, 2015) 39.

current generations. In many cases it will be right to address the needs of current generations first. However, in the context of a legal claim, these difficult choices become a barrier to establishing the legal responsibility of the state, since obligations to current generations can be used as an effective defence. Without an accepted concept of intergenerational responsibility in international human rights law, future generations' cases will remain difficult to win.

4.5 SUBSIDIARITY: PRIORITY GIVEN TO GOVERNMENT DECISIONS

The final barrier to protecting future generations' environmental rights is that human rights courts and tribunals tend to give states wide discretion to determine their own environmental priorities. This subsidiarity adopted by international bodies is an important part of respecting state sovereignty and legitimate governmental processes, and it respects the fact that national governments are often best placed to make decisions about how to deal with limited resources and address the needs and desires of their communities. However, it does leave these decisions open to the influence of 'short-termism' associated with electoral cycles and voter preferences.[101]

In the European system, the doctrine of the margin of appreciation has emerged as a way of recognising that states must balance a range of competing demands, including between different human rights objectives. Provided that states have complied with their own domestic laws, the ECtHR will generally defer to states' own judgment about how to balance different interests. The court will not rule that there has been a violation of the ECHR, except where the interference with human rights is clearly disproportionate to any legitimate aim.[102] Given that environmental impacts frequently flow from initiatives or actions which claim a legitimate social or economic purpose, the margin of appreciation and similar forms of subsidiarity can result in environmental

[101] MacKenzie (n 1).

[102] *Handyside v the United Kingdom* [1976] (Application No 5493/72, 7 December 1976); *Hatton and Others v the United Kingdom* [2003] Eur Court HR, Application No 36022/97; *Dubetska v Ukraine* [2011] Eur Court HR 30499/03; cf *Di Sarno et al v Italy* (n 29); Ole W Pedersen, 'European Court of Human Rights and Environmental Rights' in *Elgar Encyclopedia of Environmental Law* (Edward Elgar, 2019) 463, 468; Andrew Legg, 'Different Approaches to Deference in International Human Rights Law' in *The Margin of Appreciation in International Human Rights Law* (Oxford University Press, 2012) 38–66; Hana Müllerová, 'Environment Playing Short-Handed: Margin of Appreciation in Environmental Jurisprudence of the European Court of Human Rights' (2015) 24(1) *Review of European, Comparative & International Environmental Law* 83.

harm being ruled to be justifiable.[103] And with respect to the environmental rights of future generations it can mean that states will not be expected to consider future generations' interests unless their domestic laws include such a requirement. Hilson has argued that the use of the margin of appreciation in the ECtHR has resulted in the 'proceduralisation' of environmental rights claims. The ECtHR is less able to address broad-based environmental problems that do not easily fit within the template of direct, individualised harm or 'serious, specific and imminent danger'.[104]

While other international and regional human rights bodies do not employ the margin of appreciation explicitly, they all give some degree of deference to states' own decisions and priorities when competing rights issues need to be reconciled. They might impose certain minimum standards for fundamental rights but, provided these are maintained and the human rights impact is not disproportionate, states are afforded latitude to determine how to balance different factors. If a state opts to focus on short-term objectives at the expense of future generations' rights then international bodies may be unwilling to overrule. This is likely to continue without explicit principles in international human rights law which would promote a different approach.

4.6 CONCLUSION

Several features of international human rights law prevent it from being an effective instrument for the protection of future generations' environmental rights. To begin with, future generations are not recognised as holders of rights and states therefore have no obligations to respect, protect or fulfil those rights. While there has been some evolution of our understanding of states' jurisdiction, and therefore of their human rights obligations, this has not yet extended to establish duties towards future generations.

Without being recognised as holders of rights, there are no clear pathways for legal claims to be brought on behalf of future generations. Unlike some domestic jurisdictions, most international and regional human rights systems do not allow representative or public interest claims which could protect future generations' interests. Children's rights-based litigation is expanding, especially in relation to climate change, offering some incidental benefits to future generations, but this is not in itself an adequate substitute for protecting their rights specifically.

[103] *Hatton and Others v the United Kingdom* (n 102); *Greenpeace eV and Others v Germany (Admissibility)* [2009] Eur Court HR 18215/06.

[104] Chris Hilson, 'The Visibility of Environmental Rights in the EU Legal Order: Eurolegalism in Action?' (2018) 25(11) *Journal of European Public Policy* 1589.

Even if a claim could be introduced to an international human rights body, the fact that the alleged breaches will occur in the future creates further challenges. Currently, international human rights law requires states to act on imminent threats to human rights, but where harm is more remote it is difficult to show that a state is in violation of its obligations. The fact that harm will occur in the future may in fact preclude a claim in the first place, since standing often depends on showing an actual or imminent injury. And states may be able to justify an interference with future generations' rights on the basis that it is a side-effect of legitimate action to promote the rights of current generations.

The potential for international human rights law to recognise a right to a healthy environment which is ecocentric in its framing could overcome some of these challenges. An ecocentric right could allow an injury and breach to be established based on environmental harm alone, without needing to trace that harm to direct impacts on other human rights. However, despite the recent resolutions by the Human Rights Council and General Assembly, the right to a healthy environment remains at best a non-binding, loosely defined right and has not yet become enforceable at the international level.

Consequently, international human rights law is currently unable to offer protection to future generations' environmental rights. Its impact is limited to the incidental benefits that flow to future generations from the protection of current generations' rights, especially children's. While this is not insignificant, it falls short of addressing the intergenerational injustice that is inherent in many of the world's most troubling examples of environmental rights violations. Chapter 5 will suggest some changes to IHRL which could help to overcome these barriers.

5

New Theory and Practice

A Proposed New Approach for International Human Rights Law

5.1 INTRODUCTION

So far this book has attempted to show why international human rights law should protect the environmental rights of future generations, while identifying the reasons it is unable to do so in its current form. Chapter 2 explained why human rights law needs to care about future generations' environmental rights. It drew on various ways of thinking to demonstrate the moral justification for protecting future generations' environmental rights and made the case for making these reforms within international human rights law. Chapter 3 gave an overview of existing environmental rights protections and Chapter 4 explained why international human rights law (IHRL) is currently incapable of extending those protections to future generations, identifying issues of standing, admissibility, standards, causation and balancing competing rights.

Building on the analysis in earlier chapters, this chapter proposes a way forward to better protect the environmental rights of future generations. It integrates some of the key theoretical principles considered in Chapter 2 with the analysis of current gaps in protection that were identified through Chapters 3 and 4. Informed by Edith Brown Weiss's theory of intergenerational equity and the *Maastricht Principles on the Human Rights of Future Generations*, it suggests a framework of obligations to respect, protect and fulfil environmental rights for future generations. It also examines a number of existing and proposed mechanisms for protecting future generations' interests to identify options for improving the procedures for enforcing human rights.

Drawing these considerations together, the chapter argues for a number of reforms which could be made to IHRL. These include explicitly recognising the rights of future generations, or at least recognising the obligations of states to protect those rights. It also suggests creating specific pathways to enable claims to be brought on behalf of future generations, having regard to a number of models that can be found in domestic jurisdictions.

The chapter also presents one of the key contributions of the book in the form of a theory of intergenerational responsibility that can be applied to environmental rights matters to achieve a more just balance of rights and duties. This theory aims to promote fair treatment for future generations, drawing on variations of intergenerational equity and intergenerational justice, along with fundamental human rights principles of equality, universality and non-discrimination. The resultant theoretical approach also includes principles for navigating some of the legal issues identified in Chapter 4 relating to the nature of states' preventative and facilitative obligations and the appropriate way to manage risk and uncertainty and to balance competing interests. The outcome is a new theory and practice for protecting the environmental rights of future generations. Subsequent chapters then apply that approach to a number of case studies to demonstrate how it could be used by law- and policy-makers, courts and tribunals.

5.2 SUMMARY OF LIMITATIONS AND CRITERIA FOR A NEW APPROACH

As the earlier chapters demonstrated, there are some key reasons why international human rights law is currently unable to protect future generations' environmental rights adequately. As was argued in Chapter 2, an adequate legal framework would be one which delivers intergenerational justice by protecting the environmental rights of both current and future generations in a way which is fair. This section summarises and integrates the findings of earlier chapters to highlight the criteria or conditions that need to be met in devising a new approach.

Chapter 2 argued that international human rights law should align with notions of intergenerational responsibility. The many variations of intergenerational equity and justice, planetary trusts, global commons, and Indigenous peoples' notions of custodial ethics and collective continuance are all oriented towards a sense of moral obligation to care for future generations' interests and rights. IHRL is a body of law built on principles of equality, non-discrimination and justice, and these concepts should similarly extend to future people. The current evolution which IHRL is undergoing to make it better suited to protecting environmental rights (as outlined in Chapter 3) should therefore do so in a way which is fair for future generations by ensuring their rights are also adequately addressed.

Thinking about how this notion should be put into practice in reforming international human rights law, a few key criteria emerge. IHRL needs to produce fair outcomes for future generations by:

1. Conferring equal value to their rights;
2. Avoiding discrimination or discounting based purely on their place in time;
3. Understanding that fairly distributing benefits and burdens across generations often requires stronger actions now to avoid greater burdens in the future;
4. A more ecocentric understanding of environmental rights and greater recognition of the non-human world;
5. Understanding that the human rights challenges of the Anthropocene are happening at a planetary scale and intergenerational justice must therefore be global, as well as national and local;
6. Recognising and welcoming non-Western perspectives, especially Indigenous knowledges about nature and time.

Chapter 4 outlined several deficiencies of current IHRL with respect to the interests of future generations. A number of key limitations were identified:

1. Future generations are not recognised as holders of human rights and states do not therefore owe them duties;
2. Standing to enforce a claim is limited to 'victims', defined as identifiable people experiencing actual or imminent harm;
3. The content and standards of human rights duties are not sufficiently future-focused;
4. Rules for causation are designed for proving a connection to actual/imminent harm, not future harm;
5. States are often given discretion in striking a balance between competing interests, but their decision-making is often coloured by short-termism.

For future generations' environmental rights to be meaningfully addressed, a revised theory and practice of IHRL is needed which can meet the criteria of an intergenerationally just approach (the first list) by overcoming the limitations of the current system (the second list). Subsequently I set out the key components of a suggested approach designed to address the current limitations while still remaining compatible with the fundamental principles of human rights. The proposed changes could be implemented through an expanded interpretation by courts and tribunals, through General Comments and other advisory instruments, or through amendments to treaties or protocols. It is most likely that a combination of these methods would be needed to achieve comprehensive reform. They are presented here with the intention of prompting further discussion and pinpointing areas for future work.

5.3 A THEORY OF INTERGENERATIONAL RESPONSIBILITY FOR HUMAN RIGHTS LAW

The theory of intergenerational responsibility proposed here overlays principles of intergenerational equity (drawing on the work of Edith Brown Weiss and informed by alternative approaches) across the tripartite duties of human rights to respect, protect and fulfil. This enables an articulation of the scope and content of state obligations with respect to future generations. The definition of these duties is informed by the work of the World Future Council and the *Maastricht Principles on the Human Rights of Future Generations*. The theory also articulates the standards of performance which should apply to states' duties, and the circumstances in which a violation of rights could be said to have occurred. By integrating intergenerational equity into human rights law, the theory also offers a way of countering the short-termism which often influences state decision-making and judicial determinations of whether states have struck a fair balance between competing human rights.

5.3.1 *Future Generations as Rights-Holders*

In order for IHRL to protect the rights of future generations, it first needs to acknowledge that future generations are holders of rights. This might seem like an obvious, even redundant, requirement, but not everyone agrees that it is necessary. John Knox, former United Nations Special Rapporteur on human rights and the environment, argued that enabling children to bring climate litigation, for example, is an adequate substitute for protecting future generations' rights specifically, because future generations would indirectly benefit from any resulting changes to law and policy.[1] Another approach would be to focus on the content and standards of states' duties and ensure

[1] John H Knox, 'Constructing the Human Right to a Healthy Environment' (2020) 16(1) *Annual Review of Law and Social Science* 79.

that these are informed by concepts of sustainability and prevention. In this way, the question of whether future generations are rights-holders could be avoided, but they would indirectly benefit from more future-oriented laws and policies.

There is something to be said for focusing on duties and responsibilities rather than rights-ownership, given the debates that have surrounded the question of whether future generations can have rights. Human rights law arguably has put too much emphasis on rights-holders, when the key modality for advancing human rights is recognising and ensuring the performance of corresponding duties. The traditional state-centric, adversarial model of human rights litigation draws attention away from state responsibilities and puts the onus on individuals or specific groups to hold governments accountable. And there is an unavoidable difficulty in defining future generations and identifying their needs, given that each generation overlaps with the next and contains its own inequities. It may well be better to focus on states as duty-bearers and to define their duties in ways which better accommodate the protection and promotion of future generations' rights, without insisting on future generations as holders of rights.

But there are good reasons why members of future generations should be recognised as holders of rights (both individually and collectively) and why these rights should be legally enforceable. For one thing, it is appropriate because future generations are equal in human dignity to all people living now or at any other time. Recognising their rights is consistent with the fundamental tenets of human rights law, especially equality and universality. Failing to do so undermines the legitimacy of human rights law as a whole, as it entertains a form of discrimination that cannot be justified.

By ensuring that the law accords equal weight to future generations' rights, those rights are less likely to be postponed in favour of current generations. In considering how to address environmental risks, there will inevitably be instances where the rights of current and future generations are in competition with each other. As will be discussed later, the question of how to balance these competing rights can be challenging and needs careful consideration. That balancing act is easier, however, if we at least recognise that different generations of people are equally entitled to rights, rather than privileging only some as 'rights-holders'. Formally recognising future generations as holders of rights therefore has important implications for the way that those rights can be protected in practice.

The Maastricht Principles, which seek to articulate the current state of international law, provide additional support for the proposition that future generations should be recognised as holders of rights. The principles begin by stating that future generations are legally entitled to human rights.[2] The reasons for this include the fact that current human rights law recognises the rights of all people without limiting that recognition to present generations. It also points to the various laws and belief systems around the world which include responsibilities to future generations, for example, through the concept of intergenerational equity or other principles or values.

The issue of how the rights of future generations can be legally enforced is discussed in more detail subsequently, particularly in relation to the issue of standing to bring a legal claim. That is a practical matter which can be resolved. Questions of how to define state obligations with respect to future generations' rights must also be addressed. Neither of these issues detracts from the reality that future generations of people are equally entitled to human rights, and IHRL must therefore recognise them as holders of rights. As it currently stands, IHRL describes rights-holders as the people to whom states owe duties, so we must therefore consider how to adjust state duties to ensure they encompass future generations.

5.3.2 *Scope of Future Generations' Rights*

Before turning to the nature of obligations that sit with duty-bearers, we should consider which rights we say future generations possess (or will possess), particularly in relation to environmental rights. Thinking back to the discussion of the justifications for recognising future generations' rights in Chapter 2, some scholars have attempted to describe the rights which future generations possess. Edith Brown Weiss describes a set of planetary rights which flow from obligations of intergenerational equity. Importantly, planetary rights are collective rights enjoyed by each generation, and consist of 'the right to receive the planet in no worse condition than that of the previous generation, to inherit comparable diversity in the natural and cultural resource bases, and to have equitable access to the use and benefits of the legacy'.[3]

We should go further than this, however, and recognise that future generations will be entitled to the full range of rights guaranteed in international human rights law. Indeed, this would be a corollary of invoking equality and universality as the basis for recognising future generations' rights in the first place – future people must be equally entitled to the same rights as all people.

[2] 'Maastricht Principles on the Human Rights of Future Generations' art 2 <www.ohchr.org/sites/default/files/documents/new-york/events/hr75-future-generations/Maastricht-Principles-on-The-Human-Rights-of-Future-Generations.pdf>.

[3] Edith Brown Weiss, *In Fairness to Future Generations: International Law, Common Patrimony and Intergenerational Equity* (Transnational, 1989) 95.

To the extent that IHRL recognises environmental rights, those rights belong to future generations as well. As outlined in Chapter 3, environmental rights entail the environmental dimensions of all other human rights. In some jurisdictions, they also include the right to a healthy environment, though this is not yet recognised within IHRL.

As noted in Chapter 2, one of the criticisms against recognising the rights of future generations is based on the fact that those people do not yet exist and their identities are not yet known. The reasons why this criticism should fail are explained in Chapter 2, but it points to the question of whether the rights we might recognise for future generations can be individual rights or merely collective rights held by future people in general. Traditionally, most rights protected under international law are understood as individual rights – they are claims that each individual citizen has against their government. The chief exception to this has been the right to self-determination, which is enjoyed by peoples collectively and was recognised at the inception of modern international human rights as part of the United Nations' commitment to decolonisation. In more recent times, cultural rights have emerged as another group of collective rights, especially those articulated in the *United Nations Declaration on the Rights of Indigenous Peoples*.

As explained in Chapter 3, environmental rights can be both individual and collective. Some are individual rights (like the right to health) while others are group rights (like the right to culture). But the enjoyment of individual rights can have a collective dimension, and vice versa. The right to a healthy environment must be understood to have characteristics of both categories, and many of the domestic laws which recognise the right to a healthy environment define this in collective terms.

Because many of our ideas of intergenerational justice are based on concepts like the common good or a planetary trust, it makes sense to think of future generations' rights as collective rights. According to Brown Weiss, because these rights relate to planetary conditions, they are understood and enjoyed at the collective level by each generation as a whole, rather than individually. 'Of necessity these rights involve common patrimony, such as air, which cannot be parcelled out to individuals, but they are defined interests of each generation which must be protected.'[4]

However, she notes that planetary rights share a common focus with traditional, individualised human rights in that they aim to safeguard 'the welfare and dignity of human beings'.[5] And, while the enforcement of future

4 Ibid 99.

5 Ibid 114.

generations' rights must practically be done on behalf of a collective (since the individual people do not yet exist), we can still accept that each future individual will be entitled to the full set of environmental rights on an equal basis to any person living today. The fact that enforcement must be done in some generalised way does not limit the content of the rights which we recognise for future generations.

The content of rights for future generations can therefore be described as the full set of human rights, including all environmental rights, that any person today would be entitled to. These are both individual rights – the things needed for each individual person to live a life of dignity – and collective rights – those shared goods which peoples and groups enjoy in communion with each other. Future people will of course be individuals, but they will also be members of groups, peoples and communities and must therefore be entitled to both individual and collective rights.

5.3.3 *States as Duty-Bearers*

As identified in Chapter 4, a key reason why future generations are not recognised as rights-holders is because they do not fall within the jurisdiction or control of states. A crucial step in protecting future generations' rights must therefore be the recognition of state duties towards future people. This could be done by adapting the approach to extraterritorial obligations that has emerged from recent human rights cases. The elements of the approach were discussed in Chapter 4 and are heavily influenced by the Inter-American Court of Human 'Rights' *Advisory Opinion on the Environment and Human Rights*, as endorsed by the Committee on the Rights of the Child in the *Sacchi* complaint.[6]

For extraterritorial human rights impacts, jurisdiction (and therefore obligations) exists where:

1. The state has exercised effective control over activities – for example, by regulating industry or actively making policy decisions;
2. There is a causal link between the activities and negative human rights impacts; and
3. The negative human rights impact was foreseeable.

6 *Environment and Human Rights: State Obligations in Relation to the Environment in the Context of the Protection and Guarantee of the Rights to Life and to Personal Integrity (Advisory Opinion)* [2017] Inter-American Court of Human Rights OC-23/17, 23; *Sacchi et al v Argentina* [2021] Committee on the Rights of the Child Communication No 104/2019.

To convert this to a basis for intertemporal jurisdiction and duties, some adjustment of elements 2 and 3 is required. The criteria could instead be described as:

1. The state exercises control over activities (usually because they fall within the scope of the state's legislative or executive power);
2. Those activities may cause an impact on the human rights of future people (based on an objective, evidence-based risk assessment); and
3. The human rights impact is foreseeable to the state at the time it exercises control (assessed objectively).

Because this test is focused on future impacts, there will inevitably be an overlap between the second and third elements, since the evidence required to show a causal connection (element 2) will likely be related to that needed to establish foreseeability (element 3). But they can and should be identified and applied as separate elements. The second element is oriented towards the nature of impact that is anticipated, showing the causal link between the state's act or omission and the resulting effect (as opposed to an effect that might occur regardless of the state's actions). The third element focuses on whether the state has foreseen or ought to have foreseen the harm. The use of 'may' in the second element is intended to avoid stipulating any threshold of probability – again, because its purpose is to establish the causal connection between the exercise of state control and the human rights impact. The degree of likelihood will be more relevant to later considerations about the content of state duties and what was required from the state in order to address the risk of future harm. In combination, these elements establish that future people fall within the jurisdiction of the state for the purposes of establishing human rights duties, as they will be foreseeably affected by actions under the state's control.

The Maastricht Principles adopt a similar framing for the scope of states' jurisdiction, though arguably with a slightly lower threshold for when a duty will exist. Principle 14 explains that duties will exist whenever a state exercises authority or effective control, or where its conduct brings about foreseeable impacts on the enjoyment of human rights for future generations, or where the state is in a position 'to exercise decisive influence, or to take measures to realise the human rights of future generations'.[7] This general statement of the circumstances in which states owe obligations is supplemented by more specific detail on the content of states' duties to respect, protect and fulfil human rights (discussed in 5.3.4).

7 'Maastricht Principles on the Human Rights of Future Generations' (n 2) 14.

The three-step proposal presented here is also consistent with the approach taken by the Committee on the Rights of the Child in its recent General Comment on children's rights and the environment. The General Comment says that states have a 'responsibility for foreseeable environment-related threats arising as a result of their acts or omissions now, the full implications of which may not manifest for years or even decades'.[8]

5.3.4 A *Typology of States' Duties*

So far, we've laid an essential foundation by recognising that future generations have human rights and that governments owe duties with respect to those rights. In order to make those rights meaningful, however, it's necessary to give more detail to the scope and nature of those duties. Only then can we articulate more precisely what a state is expected to do and, importantly, identify when it has failed to meet those expectations.

International human rights law typically gives content to states' duties using a tripartite framework of the obligations to respect, protect and fulfil.[9] The duty to respect human rights entails an obligation to refrain from acting in a way that would interfere with the enjoyment of human rights. It is therefore construed as a negative duty.[10] The duty to protect human rights requires states to take measures to protect human rights against interference by private actors.[11] The third level of duty, the duty to fulfil human rights, is the highest level of obligation and requires that governments take positive steps to ensure that individuals and groups can enjoy their human rights.[12] For economic, social and cultural rights in particular, the duty to fulfil human rights is normally applied in conjunction with an expectation of progressive realisation, recognising that for some states, particularly the least developed nations, full realisation of human rights requires considerable resources and will take time to achieve.[13]

8 Committee on the Rights of the Child, 'General Comment No. 26 (2023) on Children's Rights and the Environment, with a Special Focus on Climate Change' [11].

9 Sumudu Atapattu, *Human Rights Approaches to Climate Change* (Routledge, 2016) 46; Ryan Goodman, Henry J Steiner and Philip Alston, *International Human Rights in Context*, 3rd ed (Oxford University Press, 2007) 185–189.

10 Goodman et al (n 9) 187.

11 Manfred Nowak, *U.N. Covenant on Civil and Political Rights: CCPR Commentary*, 2nd rev ed (NP Engel, 2005) 37–39.

12 Goodman et al (n 9) 118.

13 *International Covenant on Economic, Social and Cultural Rights* 1966 art 2 ('ICESCR'). Economic, social and cultural rights include rights like the right to education, an adequate standard of living, and the highest attainable standard of health.

We can use the tripartite duties as a starting point to develop a framework of obligations owed to future generations. Each of the three levels of duties must be fleshed out having regard to the concepts of intergenerational responsibility (discussed in more detail in Chapter 2) and considering the nature of environmental rights and what they require (covered in Chapter 3 and Section 5.3.2). In determining the content of duties, we are building on the discussion in Section 5.3.3 about the scope of a state's jurisdiction, which sets the conditions for when duties will be owed. As suggested earlier, states' duties will be triggered where an exercise of effective control is likely to impact on the rights of future generations, in circumstances where that effect was foreseeable.

Table 5.1 summarises a suggested framework of duties for protecting the environmental rights of future generations. It uses the tripartite duties as a starting point, overlaid with Edith Brown Weiss's theory of conservation of options, quality and access. It refers to 'environmental goods', being the broad category of natural systems, resources, biodiversity, landscapes etc which relate to environmental rights.

TABLE 5.1 *A framework of duties towards future generation*

Conservation of:	Options	Quality	Access
Duty to Respect	Refrain from actions which would limit the diversity of environmental goods or restrict future generations' options for enjoying environmental rights.	Refrain from actions which would pollute, degrade, destroy or otherwise diminish the quality of the environmental goods enjoyed by future generations.	Refrain from actions which would block future generations from accessing environmental goods.
Duty to Protect	Regulate other actors to preserve the diversity of environmental goods and related options for future generations.	Regulate other actors to preserve the quality of environmental goods.	Regulate other actors to protect access to environmental goods.
Duty to Fulfil	Take positive steps to maintain, restore and improve options for future generations in relation to environmental goods.	Take positive steps to maintain, restore and improve the quality of environmental goods.	Take positive steps to maintain, restore and improve access to environmental goods for future generations.

It expands on Brown Weiss's notion of conservation to include some obligations of improvement and restoration, in recognition of the fact that the full enjoyment of environmental rights requires more than simply conservation of resource and that worsening environmental conditions must be addressed to avoid creating unfair burdens for future generations. The specific requirements of each duty will of course vary depending on the circumstances, and there will be times when an impact on future generations' rights is justifiable – for example, because it is necessary to protect or fulfil other human rights. The following discussion gives some more detail about how this could be determined.

Duty to Respect

Considered with respect to future generations, the duty to respect human rights encompasses a duty not to act now in a way which would prevent people from being able to enjoy their environmental rights at a future time. It applies when the impact of the state's act or omission could have a foreseeable negative effect on the enjoyment of human rights. The Maastricht Principles describe the duty to respect in these terms: 'States must refrain from conduct they foresee, or ought reasonably to foresee, will create or contribute to a substantial risk of violations of the human rights of future generations'.[14]

Using a broad understanding of environmental rights, this requires not only conserving natural resources by using them sustainably, but also preserving the quality and diversity of all environmental goods. It further requires states to refrain from directly or indirectly damaging the environment in ways that create an undue burden for future generations.

To use climate change as an illustration, the duty to respect human rights would require states to refuse new fossil fuel developments on the basis that their greenhouse gas output would lead to negative human rights consequences for future people. It would also demand that states avoid any mitigation or adaptation action that would be likely to have a negative impact on the rights of future generations.[15] The Maastricht Principles give some other examples of conduct that could be considered a violation of the duty to respect, such as polluting or degrading ecosystems, unsustainably depleting natural resources, contributing to biodiversity decline, or impairing the ability of future generations to respond to environmental problems in the future.[16]

[14] 'Maastricht Principles on the Human Rights of Future Generations' (n 2) 16.

[15] See Bridget Lewis, 'The Rights of Future Generations within the Post-Paris Climate Regime' (2018) 7(1) *Transnational Environmental Law* 12–13.

[16] 'Maastricht Principles on the Human Rights of Future Generations' (n 2) 17.

Duty to Protect

The duty to protect requires states to take steps to prevent corporations and other private actors from impacting negatively on future generations' rights, or at least to minimise the negative future consequences of their activities. In order to do this, states must be able to identify circumstances in which a threat to future generations' rights might arise, and should take steps to prevent or mitigate those circumstances. Adequate environmental regulations must be in place to protect against degradation, destruction or pollution of the environment and loss of biodiversity.

The Maastricht formulation says that states 'must take all necessary measures to protect the human rights of future generations against substantial risks posed by the conduct of public and private actors, including business enterprises'. These necessary measures include adopting and implementing legislation, regulations, procedures and institutions which allow threats to be identified and appropriate preventative action to be taken. They also include ensuring that effective and accessible judicial or other remedial avenues are available when a violation of rights occurs.[17]

Again, to consider the example of climate change, the duty to protect requires states to phase out the use of fossil fuels by regulating the mining and energy sectors. Wherever a state delegates its mitigation or adaptation obligations to a private actor, it must ensure that the same standards of human rights protection are maintained and enforced. The Maastricht Principles specifically note the need to regulate scientific research and activities that pose a foreseeable risk to future generations, including geoengineering, and the need to prevent monopolisation or other 'abusive corporate control of data' that might undermine the rights of future generations.[18]

Duty to Fulfil

The duty to fulfil is perhaps the most difficult duty to conceive and deliver for future generations, since it implies anticipating their needs and taking preemptive steps. It's one of the reasons some scholars have argued against recognising future generations' rights, since it is so difficult to predict what those rights will require. However, if we accept that our actions today will influence future generations' enjoyment of rights, then we can start to identify some actions we should be taking to maximise that enjoyment.

The Maastricht Principles frame the obligation to fulfil human rights as a duty to create an enabling environment in order to address poverty and

[17] Ibid 18.
[18] Ibid 19.

inequality, and to provide adequate financial and other resources. The measures states must take include recognising future generations' rights in legislation and embedding obligations within governmental frameworks, including through creating dedicated agencies and processes that support the fulfilment of their rights. States must ensure that appropriate risk assessments are conducted whenever activities or legislation might have an impact on future generations.[19]

These measures are designed to create conditions that put future generations in the best possible position to enjoy their human rights. By their very nature, they entail the advancement of human rights for present generations as well. The Maastricht Principles note that a failure to ensure 'at the very least, the satisfaction of essential levels of social, economic and cultural rights for present generations, and to take measures that enable future generations to ensure these levels for themselves' would be a violation of the duty to fulfil.[20]

With respect to the environment, the duty to fulfil goes beyond merely conserving natural resources or using them sustainably. It requires positive action to shift away from practices that create waste and pollution and move towards more sustainable means of production and consumption. We should also take steps to restore and regenerate landscapes and ecosystems so that future generations are not left with unreasonable burdens to rehabilitate what we have damaged and depleted.

Using our example of climate change, the duty to fulfil reinforces the fact that states must act quickly to address the future consequences of climate change.[21] Wealthy states in particular must rapidly transition away from fossil fuels – the longer this action is delayed the worse the burdens will be for future generations. Positive steps must also be taken to establish long-term adaptation measures and build capacity to support future generations in dealing with climate change. Where climate change impacts are predicted with a high degree of certainty, states should be taking measures to ensure that this impact is minimised wherever possible, including through taking actions to enhance infrastructure, systems and capacity.[22]

Taken together, the three levels of duty establish a framework for articulating what is required from states in order to safeguard the rights of future generations. They go beyond merely asking states to refrain from making

[19] Ibid 20.

[20] Ibid 21.

[21] Daniel Bodansky, 'Climate Change and Human Rights: Unpacking the Issues' (2014) 38(3) *Georgia Journal of International & Comparative Law* 511, 520.

[22] Bridget Lewis, 'Human Rights Duties towards Future Generations and the Potential for Achieving Climate Justice' (2016) 34(3) *Netherlands Quarterly of Human Rights* 218.

things worse for future generations and ask that they take positive steps to create a better world for them.

5.4 PRINCIPLES FOR LITIGATING FUTURE GENERATIONS' RIGHTS

Recognising rights and duties and defining their respective content and standards is an essential part of expanding IHRL to accommodate future generations' environmental rights, but it's not enough on its own to ensure that those rights can be adequately protected. IHRL still needs to provide some way for those rights and duties to be enforced. Chapter 4 identified a number of reasons why IHRL is currently unable to deal with claims on behalf of future generations. First and foremost is the fact that future generations are not considered to be 'victims' who have standing to bring a claim, and currently there is no option for representative or public interest claims which could litigate their interests. Other issues also relate to standards of performance and rules of causation which currently hinder successful claims based on anticipated future harm. These are complicated by the reality that states must balance future generations' rights with potentially competing obligations to meet the needs of current generations.

This section begins by examining some of the options for standing, taking inspiration from some of the approaches that have been implemented in different domestic jurisdiction. It identifies several principles which should inform the performance of state duties, having regard to the need to balance responsibilities now and into the future. Together, these principles could enable future generations' rights to be litigated within IHRL so that violations can be established and state duties enforced.

5.4.1 *Standing to Bring a Claim*

Earlier in this chapter, it was argued that international law could recognise future generations as holders of rights and states as duty-bearers with respect to those rights by expanding the understanding of state jurisdiction, similar to the reasoning of the Committee on the Rights of the Child in the *Sacchi* decision[23] and the Inter-American Court's Advisory Opinion.[24] To accompany this reframed concept of jurisdiction and duties, international human rights

[23] *Sacchi et al v Argentina* (n 6).

[24] *Environment and Human Rights: State Obligations in Relation to the Environment in the Context of the Protection and Guarantee of the Rights to Life and to Personal Integrity (Advisory Opinion)* (n 6).

law needs to amend rules of standing to allow claims for foreseeable future harm. As was explained in Chapter 4, international human rights bodies typically require a person or group to be a victim of an actual or imminent violation of rights in order for their claim to be admissible. That is, they must show 'either that an act or an omission of a State party has already adversely affected their enjoyment of such right, or that such an effect is imminent'.[25] This narrow understanding of harm and victimhood rules out claims on behalf of future generations. To resolve this, another pathway needs to be opened up to bring their claims before courts and tribunals so that states' obligations can be meaningfully enforced. A number of options for standing are considered here, informed by practice and procedures in different jurisdictions.

The first possible reform is to change the definition of 'victim' for the purposes of human rights litigation to include future generations. Recent and relevant soft law instruments have proposed this kind of expansion. The Maastricht Principles define 'victims' to include 'persons, groups and peoples, who face a substantial and reasonably foreseeable risk of suffering human rights violations'.[26] The *Strasbourg Principles on Environmental Rights* recognise standing for both victims and 'potential victims', with the latter referring to 'persons to whom the violation would cause harm in a foreseeable and not too distant future'.[27] The Strasbourg Principles also note that 'the term victim ought to evolve in the light of conditions in contemporary society and be applied without excessive formalism'.[28]

Drawing on these ideas together with the expanded concept of state jurisdiction, a new definition of 'victim' could include a person, group or people who will be foreseeably harmed by actions under a state's control. As long as there are reasonable grounds for reaching this conclusion, the future victims would satisfy the threshold for standing to bring a claim against the state.

Obviously, however, future generations can't bring these claims themselves; some form of representation is needed to operationalise standing for future generations. One option is for children or young people to include future generations in litigation that addresses long-term environmental or climate harm. At the domestic level, cases which have successfully considered intergenerational equity have mostly been brought by children and young people.

[25] *EW et al v The Netherlands* [1993] UN Doc CCPR/C/47/D/429/1990 (1993) Human Rights Committee Communication No 429/1990, para 6.4.

[26] 'Maastricht Principles on the Human Rights of Future Generations' (n 2) Principle 29.

[27] 'The Strasbourg Principles of International Environmental Human Rights Law – 2022' (2022) 13 *Journal of Human Rights and the Environment* 195, 12.

[28] Ibid Principle 11.

For example, cases have been successful in Colombia, Germany and Australia, and are underway in Canada and the United States, among other places.[29]

In Australia, a group of young people challenged the Queensland government's decision to issue a coal mining lease on the basis that the climate impacts which would result from the eventual burning of the coal would violate their rights under the *Human Rights Act 2019 (Qld)*.[30] The Land Court had regard to intergenerational equity in weighing up the likely impact of the mine on human rights. President Kingham of the Land Court emphasised the future impacts of climate change on the rights of children, pointing out the need to protect children from both the increasing risks of climate change and the burdens of dealing with adaptation and mitigation in the future.[31] She also highlighted the intergenerational dimensions of Aboriginal and Torres Strait Islander peoples' rights to culture, which include the right (and responsibility) to transmit cultural practices and language to future generations and to conserve natural resources and sacred sites for the benefit of their descendants.[32] While intergenerational responsibility was therefore influential in the decision, the rights considered were those of living children and First Nations peoples, not the rights of future generations per se.

In the Canadian province of Québec, a group of young people have commenced legal action against the Canadian Government alleging that, by adopting climate policies that it knows to be inadequate, the government is violating their rights to life and to inviolability and security of the person under the Canadian Charter and the Québec Charter.[33] Further, they argue that these inadequate targets violate their right to live in a healthful environment, guaranteed in the Québec Charter.[34]

These cases implicitly protect the environmental rights of future generations by addressing predicted climate harms and therefore have potential to

29 *Demanda Generaciones Futuras v Minambiente* [2018] (11001-22-03-000-2018-00319-01, 5 April 2018); *Neubauer et al v Germany* [2021] Federal Constitutional Court 1 BvR 2656/18, 1 BvR 78/20, 1 BvR 96/20, 1 BvR 288/20; *Waratah Coal Pty Ltd v Youth Verdict Ltd and Others (No 6)* [2022] Land Court of Queensland QLC, 21, 1339; *ENVironnement JEUnesse v Canada* [2021] QCCA Quebec Court of Appeal 500-09-028523-199, 1871; *Juliana v US* [2016] US District Court (District of Oregon, Eugene Division) Case No 6:15-cv-01517-TC.

30 *Waratah Coal Pty Ltd v Youth Verdict Ltd and Others (No 6)* (n 29).

31 Ibid 1602.

32 Ibid 1563.

33 *Canadian Charter of Rights and Freedoms* (1982) s 7; *Québec Charter of Human Rights and Freedoms* (1975) s 2; *ENvironment JEUnesse v Procureur General du Canada (Application)* (Superior Court of Québec, 26 November 2018); *ENvironment JEUnesse v Procureur General du Canada*, No 500-06-000955–183 (Superior Court of Québec, 11 July 2019).

34 *ENvironment JEUnesse v Procureur General du Canada (Application)* para 2.88 ff. These cases will be discussed in more detail in Chapter 8.

achieve some level of judicial recognition of obligations towards them. However, for the reasons discussed in Chapter 4, it is not adequate or appropriate to rely on children's litigation to advocate for and enforce the rights of future generations. A different pathway is needed which can consider future generations' rights independently.

Another option for representation of future generations' rights would be to allow public interest claims (or actio popularis) within international human rights courts and committees.[35] There are examples of this in domestic jurisdictions, many of which make use of special constitutional provisions that allow legal enforcement of constitutional rights. For example, the constitution of Ecuador allows standing for any person to enforce the environmental rights found in its constitution, including the rights of nature itself.[36] Kenya's constitution specifies that any person may bring an action to enforce the rights included therein, either on their own behalf, on behalf of another person, or in the public interest.[37]

In some jurisdictions, constitutional enforcement provisions have been utilised to address future generations' rights explicitly. One example of this is *Demanda Generaciones Futuras v Minambiente* (*Future Generations v Ministry of Environment*), a decision of the Colombian Supreme Court.[38] The case was initiated through the *acción de tutela* process, a special form of legal action established within the Colombian Constitution.[39] The *tutela* is a form of *amparo* complaint common in Latin American states, which allows individuals to bring an action quickly and cheaply when they feel fundamental rights under the Constitution have been breached. While the complaint was commenced by a group of young people, the Colombian Supreme Court recognised the environmental rights of future generations based on ethical duties of solidarity and the intrinsic value of nature. In a judgment written by Judge Luis Armando Tolosa Villabona, the court stated that:

> The protection of fundamental rights not only involves the individual but implicates the 'other'. The neighbour is otherness; its essence, the other people that inhabit the planet, also includes other animal and plant species. But in addition, this includes the unborn, who also deserve to enjoy the same environmental conditions that we have.[40]

35 Brown Weiss, *In Fairness to Future Generations* (n 3) 123.

36 *Constitution of Ecuador 2008* arts 11, 86.

37 *Constitution of Kenya 2010* art 22.

38 *Demanda Generaciones Futuras v Minambiente* (n 29).

39 Marcia W Coward, Peter B Heller and Anna I Vellve Torras, 'Colombia's Constitution of 1991 with Amendments through 2005' 130, art 86.

40 *Demanda Generaciones Futuras v Minambiente* (n 29) 5.2.

Under the Philippines Constitution, any citizen may bring a claim to enforce environmental rights and obligations. In the case of *Minors Oposa v Factoran*, the Supreme Court of the Philippines held that a group of young plaintiffs had standing to bring a claim on behalf of future generations as a corollary of the right to a healthy environment.[41] The children were recognised not only as having a right to a healthy environment, but also corresponding duties to ensure the same for others, including future generations. As a result, the court granted them standing to pursue the claim on behalf of future generations as well, on the basis that doing so was part of fulfilling their own intergenerational obligations.[42] Since this case, the Philippines Supreme Court has introduced procedural rules to support any citizen in bringing a claim to enforce environmental rights under the Constitution, including on behalf of minors and future generations.[43] The rules waive filing fees and other procedural barriers to facilitate public interest claims in support of the environment, without the need to show a direct individual impact.

In the African regional human rights system, the Commission on Human and Peoples' Rights may consider actio popularis from individuals and organisations without the need for them to show a personal injury, or even that they 'know or have any relationship with the victim'.[44] This enables a much broader range of matters to come before the Commission, and arguably a more meaningful enforcement of environmental rights.[45] However, the African Commission is somewhat unusual among the regional human rights bodies. As noted in Chapter 4, public interest claims or claims 'in abstracto' are generally not permitted. Opening up standing to public interest claims may generate concerns that cases could be overly broad or speculative. However, this can be mitigated by insisting that, while the case might be brought by a non-victim in the public interest, there are still reasonable

[41] *Minors Oposa v Secretary of the Department of Environment and Natural Resources* (1994) 33 ILM 173.

[42] Ibid; Ted Allen, 'Note. The Philippine Children's Case: Recognizing Legal Standing for Future Generations' (1993) 6(3) *Georgetown International Environmental Law Review* 713; Sumudu Atapattu, 'Intergenerational Equity and Children's Rights: The Role of Sustainable Development and Justice' in Claire Fenton-Glynn (ed), *Children's Rights and Sustainable Development: Interpreting the UNCRC for Future Generations* (Cambridge University Press, 2019) 167.

[43] Constitution: constitutional right of the people to a balanced and healthful ecology.

[44] *African Charter on Human and Peoples' Rights* 1981 art 55; *Article 19 v Eritrea* (2007) 73 AHRLR African Commission on Human and Peoples' Rights Communication 275/2003; *Law Society of Zimbabwe and Others v Zimbabwe* [2016] African Commission on Human and Peoples' Rights Communication 321/2006.

[45] Elinor Buys and Bridget Lewis, 'Environmental Protection through European and African Human Rights Frameworks' (2022) 26(6) *International Journal of Human Rights* 949.

grounds to foresee that there will be future victims, in line with the definition outlined earlier.

An alternative, somewhat narrower option would be to permit claims by an individual or organisation who represents a defined group of future people. Standing would be based on the foreseeable impact of a state act or omission on future people who share some common characteristic that gives rise to their victim status – for example, people who will live in proximity to a proposed development or share a common health issue. In contrast to an actio popularis claim, which is based on the public interest more broadly, this sort of case would be based on the standing of the defined group of future people, similar to a class action.

The case could proceed through a representative, such as a litigation guardian or guardian ad litem, who could be appointed to represent the specific interests of future generations.[46] A litigation guardian or friend is usually appointed for a litigant to help make decisions about their case in circumstances where they are unable to participate in proceedings independently and require some support (e.g. a child in a family law matter or an adult who lacks capacity). For future generations, the identity of victims would necessarily be unknown, but the class of victims could be defined by reference to known criteria. A version of this occurred in *Juliana v United States*, where young people brought a case against the United States Government in their own names, but Dr James Hansen was appointed as guardian *ad litem* to represent future generations.[47] This recognises the reality that the interests of children and young people will not necessarily align with the interests of unborn generations, and courts may need to consider these separately. However, we have also seen that courts are not always willing to consider these more remote issues, and may exclude arguments which exclusively relate to future generations.[48]

Instead of expanding the concept of standing to allow complaints by individuals or organisations who seek to represent future generations, another option would be to create a permanent body or office which would have standing to bring a complaint to the existing international human rights bodies

[46] Brown Weiss, *In Fairness to Future Generations* (n 3) 123.

[47] *Juliana v US* [2020] United States Court of Appeals for the Ninth Circuit 18-36082; *Juliana v US* (n 29).

[48] See e.g. *State of the Netherlands v Urgenda Foundation* [2018] Hague Court of Appeal ECLI:NL:RBDHA:2015:7145; *State of the Netherlands v Urgenda Foundation* (2019) 19/00135 Supreme Court of The Netherlands ECLI:NL:RBDHA:2015:7145. Both the Court of Appeal and Supreme Court avoided consideration of future generations' claims, concluding that it was sufficient to address climate harms facing the current generation.

in cases of foreseeable future human rights violations. This could be modelled on an ombudsman or commissioner for future generations, like those established in domestic jurisdictions. For example, in Hungary, the Ombudsman for Fundamental Rights has responsibility for overseeing compliance with constitutional rights and responsibilities. This office is home to a Deputy Commissioner dedicated to protecting the interests of future generations. The Ombudsman and associated officers have powers to conduct inquiries, refer matters to the Constitutional Court and propose laws in the interests of future generations.[49]

In Wales, the Commissioner for Future Generations makes recommendations to government on how to meet the well-being goals set out in the *Well-being of Future Generations (Wales) Act 2015*. These include goals relating to sustainability, health, equality, culture and language. The Commissioner has power to review how well a public body is performing its obligations under the legislation. Several other countries have established mechanisms or institutions to advance the interests of future generations in some form, though their functions are mostly supervisory or advisory. Significantly, they do not have standing within the regular legal systems to litigate on behalf of future generations.

The idea of an international body representing future generations isn't a new one. Edith Brown Weiss advocated for the creation of an international ombudsman for future generations whose functions could include supervising the implementation of intergenerational obligations, investigating alleged breaches and advocating for intergenerational equity in international law-making.[50] The idea of a UN High Commissioner for Future Generations was advanced at the Rio+20 Conference on Sustainable Development in 2012.[51] More recently, the UN Secretary-General has proposed the creation of a Special Envoy for Future Generations.[52] This sort of body would be

[49] Office of the Commissioner for Fundamental Rights of Hungary, 'The Role of the Ombudsman for Future Generations', *AJBH-EN* <www.ajbh.hu/web/ajbh-en/the-role-of-the-ombudsman>; Marcel Szabó, 'National Institutions for the Protection of the Interests of Future Generations' (2015) 2(2) *E-Publica: Revista Electrónica de Direito Público* 6.

[50] Brown Weiss, *In Fairness to Future Generations* (n 3) 124–126.

[51] Rio+20 Conference on Sustainable Development, *The Future We Want (Outcome Document)* (2012) <https://sustainabledevelopment.un.org/rio20/futurewewant>; see also United Nations Secretary-General, *Intergenerational Solidarity and the Needs of Future Generations* (General Assembly 68th sess, 5 August 2013) 93 <https://sustainabledevelopment.un.org/content/documents/2006future.pdf>.

[52] António Guterres, *Our Common Agenda: Report of the Secretary-General* (2021) 45 <www.un.org/en/content/common-agenda-report/assets/pdf/Common_Agenda_Report_English.pdf>; António Guterres, *To Think and Act for Future Generations* (No Policy Brief No 1, March 2023) <www.un.org/sites/un2.un.org/files/our-common-agenda-policy-brief-future-generations-en.pdf>.

tasked with ensuring that the interests of future generations are represented in treaty negotiations and meetings of international bodies.

Having a strong voice to speak on behalf of future generations would certainly help prevent their rights and interests being devalued, postponed or ignored as a result of being overshadowed by the more visible and tangible demands of present generations. An advocate for future generations would be able to ensure that states and international organisations are equipped with the necessary information about future impacts to better enable them to make decisions which are just and balanced.

However, there are likely to be limitations to the effectiveness of an international ombudsman or commissioner for future generations. For one thing, rules of sovereignty and non-interference would constrain its ability to intervene in matters which are normally considered to fall within the authority of nation states.[53] It may also struggle to compete with the negotiating power of sovereign states where new laws and policies are being developed.[54] Without strong support from states who are willing to subject themselves to new obligations and greater scrutiny, it is likely to be limited to an advisory role without strong enforcement powers.

Ideally, a voice for future generations would be able to go beyond a merely advisory role, and would have the power to advocate on their behalf, including through possessing standing to litigate. On the particular question of standing, new rules would be needed to enable an international office to litigate on behalf of future generations within existing courts or committees. Some threshold of foreseeability of harm could still be required to ensure cases are not too speculative and have some reasonable prospect of success. For reasons of capacity, time and resources, these would likely be carefully selected cases, probably designed to test the application of IHRL to certain issues. There is therefore a risk that granting standing to an international representative organisation would create a 'gatekeeper' for litigation. At worst, valid cases might be excluded if they did not fit the priorities of the organisation – for example, because they duplicated issues already litigated or were not considered to be of high significance. For these reasons, this is not considered a preferable approach (at least not on its own). Some form of standing that can be accessed by individuals or organisations independently is required to ensure violations of future generations' rights can be litigated when necessary.

[53] Brown Weiss, *In Fairness to Future Generations* (n 3) 126.

[54] Malgosia Fitzmaurice, 'Intergenerational Equity Revisited' in Isabelle Buffard et al (eds), *International Law between Universalism and Fragmentation* (Brill Nijhoff, 2008) 226.

A final option to be considered is the possibility of states having standing to bring a case to the International Court of Justice in support of future generations' rights. The ICJ is currently preparing an Advisory Opinion on the question of the Obligations of States in Respect of Climate Change.[55] Among other questions, the ICJ has been asked to consider the legal consequences for states who have caused climate harm to 'past and future generations'. The Advisory Opinion will provide much needed clarity on the scope of states' obligations and responsibility with respect to future climate harms. However, it will not make findings in respect of particular states' actions and will not constitute a legally binding ruling. For this to occur, the ICJ's contentious jurisdiction would need to be activated via a claim by a state with standing under the rules of state responsibility or diplomatic protection.

As discussed in Chapter 4, states will have standing to bring a claim to the ICJ where they have suffered an injury themselves or where they allege a breach of an obligation erga omnes.[56] An injury for the purposes of standing includes injuries to the state's nationals.[57] For claims on behalf of future generations to be reliably admissible in the ICJ, expansion of these doctrines is required. The doctrine of diplomatic protection would need to be interpreted to include future nationals. Arguably this isn't a huge leap, given that states can already base standing on injury to their own interests, so future harm to citizens could be construed as an injury to the state itself. However, opening up standing in this way would be of little assistance where future generations face risks caused by their own state's actions.

To address this, intergenerational environmental obligations need to be recognised as *erga omnes* in nature, so that any state could take legal action to enforce them. To complete the set of tools to enforce environmental rights of future generations in the ICJ, legal obligations should be recognised as extraterritorial duties, so that actions can be taken against other states for transboundary harms.

In terms of standing at the ICJ, progress is likely to be slow and challenging. For one thing, it is difficult to see environmental rights or intergenerational equity being added to the category of erga omnes obligations. To date, the ICJ has only recognised a limited number of obligations as being erga omnes – including protections against genocide, slavery and racial discrimination[58] – and

[55] *Obligations of States in Respect of Climate Change (Request for Advisory Opinion)* (International Court of Justice, 29 March 2023).

[56] *Responsibility of States for Internationally Wrongful Acts 2001* arts 42, 48.

[57] *Draft Articles on Diplomatic Protection 2006.*

[58] *Barcelona Traction, Light and Power Company, Limited (Belgium v Spain)* [1970] ICJ Rep 3.

obligations associated with the right to self-determination.[59] While environmental rights are fundamental and may be of existential importance to future generations, they are also a wide category that includes numerous obligations (as outlined later). Only the most serious obligations are likely to rise to the level of erga omnes. Arguably, rights to the environmental means of subsistence and self-determination, or protection from life-threatening environmental pollution or degradation could be considered serious enough to attract erga omnes obligations.[60] Even with this expansion, action in the ICJ is likely to be reserved for the most egregious of cases, and will only be possible where states decide to pursue a matter, which would be a time-consuming and costly exercise. Standing in the ICJ is therefore a welcome but arguably insufficient mechanism to protect the environmental rights of all future generations.

Having regard to all these possible mechanisms for standing, the preferred options would be:

1. Expanding the concept of 'victim' within IHRL to include not just persons affected by actual or imminent harm, but anyone who will face reasonably foreseeable harm as a result of a state act or omission.
2. Expanding the rules of standing in IHRL to enable either public interest claims or representative claims managed by an approved guardian. These claims should be available in UN treaty bodies. Regional human rights courts should also ensure standing rules allow representative claims in one form or another.

To ensure environmental rights can also be enforced in the International Court of Justice, additional steps should be taken, namely:

1. Expanding the doctrine of diplomatic protection to include future nationals.
2. Recognising environmental rights that are essential to survival as giving rise to obligations erga omnes.

These changes could be achieved through a combination of actions. Amending a treaty or adopting a new protocol would offer the most certainty and legitimacy, but would require significant state support and would take time to negotiate. Many of the changes could arguably be achieved through

59 *East Timor (Portugal v Australia)* [1995] ICJ Rep 90; *Legal Consequences of the Construction of a Wall in the Occupied Palestinian Territory (Advisory Opinion)* (2004) [2004] ICJ Rep 136.

60 This could draw inspiration from the definition of ecocide, which has been defined as 'unlawful or wanton acts committed with knowledge that there is a substantial likelihood of severe and widespread or long-term damage to the environment being caused by those acts' Independent Expert Panel for the Legal Definition of Ecocide, 'Commentary and Core Text'.

interpretation by courts or committees, for example, through a General Comment of a United Nations treaty body or an Advisory Opinion. This process has already occurred to a large extent in relation to extraterritorial human rights duties. The Maastricht Principles have already demonstrated that the same can happen for future generations based on existing treaty law and custom. Further, an authoritative body like the International Law Commission could offer a statement of the law in relation to diplomatic protection and erga omnes obligations.

Changing the approach to standing to recognise future 'victims' clearly widens the class of claims that could come before courts and tribunals, leaving states open to a wider range of potential actions. However, the foreseeability requirement proposed for future victimhood will narrow potential claims to some degree. Also, consistent with the approach in international and regional human rights bodies, claimants would need to exhaust any available local remedies before a complaint will be admitted. This helps to manage the case load of international and regional bodies, and might encourage parties which don't already have a public interest avenue for human rights claims to establish one in order to avoid potential international complaints.

Further, rather than be concerned about a possible 'floodgate' opening here, it is important to remember that standing is only the first step in the process of litigation, and doesn't necessarily mean states will be found responsible for a violation of rights. The standards of performance and other principles discussed in this chapter will still apply to determine whether a state has breached its duties.

5.4.2 *Establishing a Breach of Obligations*

Once a claim has been admitted to a human rights court or committee via a representative of future generations, the next step is to show that the respondent state has violated its obligations. In order to do this, we need to be able to say with some confidence when a state should be considered to have breached its obligations. In part, this is about articulating standards of performance – how much should states be expected to do in discharging the obligations outlined earlier? We can't know precisely what vulnerabilities and capabilities future generations will have or the risks that they will face, and while we may foresee some consequences of our actions, there will always be some degree of uncertainty as to how and when future generations might be affected. Given this reality, when should human rights law view states as having done enough?

Chapter 4 noted that the emphasis on imminence in current human rights practice isn't only a limitation for establishing standing but also creates challenges for showing that a state has breached its obligations, because of the difficulty of showing that harm is likely to occur. With that in mind, what kind of harm should be viewed as a violation of human rights and how can we prove that it is likely? And in a context where we are seeking to establish state responsibility for future harm, how should we approach questions of causation across an expanded timeframe? Perhaps the most challenging question of all is how states should balance potentially competing rights, given their multidimensional obligations to different groups at different points in time. When would it be justifiable to limit the rights of either current or future generations in order to safeguard the rights of the other?

This section attempts to answer some of these questions. It draws on existing principles of human rights and other law and recommends a framework of standards and rules which would flesh out state obligations and help determine when these have been violated.

Due Diligence

There are a number of existing principles of international environmental and human rights law which can be adapted to help us with this task. The first of these is due diligence. Due diligence is commonly applied to states when considering their responsibilities to address a risk of harm.[61] It first requires an assessment of the risk associated with a proposed action, both in terms of gravity and likelihood. Once the risk of harm is established, appropriate steps must be taken to prevent or mitigate the harm. Due diligence has been a longstanding part of international environmental law, where it is a particularly important component of the duty not to cause transboundary harm.[62] In the context of transboundary harm, courts and experts have been clear that this is not a duty to prevent harm entirely, but rather a procedural duty to assess the risk and implement appropriate preventative measures. As long as these steps are taken, no violation of the duty will be found, even if the harm does in fact eventuate.[63] The International Court of Justice described this in the *Pulp Mills* judgment as an obligation of conduct, rather than one of result.[64]

61 David Boyd and Stephanie Keene, *Essential Elements of Effective and Equitable Human Rights and Environmental Due Diligence Legislation* (Policy Brief No 3, June 2022) <www.ohchr.org/sites/default/files/documents/issues/environment/srenvironment/activities/2022-07-01/20220701-sr-environment-policybriefing3.pdf>.

62 *Trail Smelter Case (USA v Canada)* (1941) III RIAA 1905; *Case Concerning Pulp Mills on the River Uruguay (Argentina v Uruguay)* [2010] ICJ Rep 14 (Separate Opinion Judge Cançado Trindade) 181.

63 International Law Commission, 'Draft Articles on the Prevention of Transboundary Harm from Hazardous Activities' 7 <http://link.springer.com/10.1007/978-1-4020-8367-9_23>.

64 *Case Concerning Pulp Mills on the River Uruguay (Argentina v Uruguay)* (n 62) 187.

Due diligence is also a common mode of transposing human rights obligations to businesses and other entities.[65] The Guiding Principles on Business and Human Rights use due diligence as the key framework for articulating the responsibilities of business entities. Businesses must have a due diligence process in place which enables them to identify actual and potential risks to human rights, and to put in place appropriate steps to prevent or mitigate those risks.[66] To ensure these duties are performed, states must incorporate them into their business regulations and support them with appropriate enforcement mechanisms. This falls within states' duty under international law to protect human rights.[67]

The utility of due diligence with respect to future generations' environmental rights lies in the fact that it imposes an obligation on states to assess risk and take appropriate preventative measures. States cannot be excused merely because they haven't investigated the future risks. Rather, they have a positive obligation to look to the future and assess any potential human rights impacts. And because we're talking about future risks, it's appropriate to apply a duty of conduct rather than of result (to use the language of the ICJ in *Pulp Mills*), because we can only judge states on whether they've taken appropriate preventative measures, not on the final consequences.

However, an adequate standard of due diligence for future environmental rights needs to allow that, in some circumstances, the risk is so great that no mitigation measures would be sufficient. Where the risk of harm is serious enough, the required approach may be to refrain from proceeding with the action at all.

Precautionary Principle

To help us understand when this threshold of serious harm applies and how to deal with it, we can look to another relevant concept, the precautionary principle. The precautionary principle is commonly found in environmental law, although definitions of it can differ. A commonly invoked definition comes from the Rio Declaration on Environment and Development, which provides that, where there is a credible threat of serious or irreversible harm to the environment, scientific uncertainty should not be used as a reason for

65 Boyd and Keene (n 61).

66 Office of the High Commissioner for Human Rights, 'Guiding Principles on Business and Human Rights: Implementing the United Nations "Protect, Respect and Remedy" Framework' paras [15], [17] <www.ohchr.org/documents/publications/GuidingprinciplesBusinesshr_eN.pdf>.

67 Ibid 1–3.

postponing action to prevent that harm.[68] A more positive formulation is sometimes used, which requires states to take appropriate precautions whenever there are risks of serious environmental harm.[69]

Applying this principle to the sorts of human rights issues we're considering here, if there is a risk that an act or decision might lead to serious or irreversible harm to the environment, states should take precautions to avoid that result. The fact that future damage is not certain should not be a reason to refrain from taking precautionary measures where those measures would safeguard rights and interests. Where preventative or precautionary steps are reasonably available, states would be expected to take those steps.[70] If there are no reasonably available measures which would prevent the harm that will flow from a proposed activity, then the activity should not proceed at all.

Causation

As discussed in Chapter 4, the need to demonstrate a link between a state's act or omission and environmental harm which impinges on human rights has proved a challenge in litigating future environmental rights cases. Courts and committees have been unwilling to conclude that a state has caused a violation while the human rights impact is not imminent. Where the harm is temporally remote, courts might consider that it is possible to intervene to prevent the harm, thus breaking the chain of causation.[71] Furthermore, in environmental rights cases, particularly those concerning climate change, the complexity of causal relationships and the variety of contributing factors have disrupted conventional approaches to causation.

The requirements of causation are closely linked to the way rights and duties are defined, however, and with a revised understanding of the content of environmental rights, it becomes significantly easier to establish causation. As explained earlier, environmental rights can be interpreted through an intergenerational lens which places much greater emphasis on duties of protection and prevention. These duties are further interpreted with reference to the principles of due diligence, precaution and prevention.

[68] United Nations Environment Programme, *Rio Declaration on Environment and Development* (No A/Conf.151/26, 14 June 1992) Principle 15.

[69] Daniel Bodansky, *Deconstructing the Precautionary Principle* (SSRN Scholarly Paper No ID 888831, Social Science Research Network, 10 March 2006) <https://papers.ssrn.com/abstract=888831>.

[70] United Nations Secretary-General (n 51) 18.

[71] See e.g. *Ioane Teitiota v New Zealand* [2019] Human Rights Committee Communication No 2728/2016.

In this context, causation should be thought of not in terms of whether a state's acts or omissions have caused the ultimate human rights impact, but rather in terms of whether those acts or omissions have addressed the risk of future harm. This aligns with the approach to jurisdiction outlined earlier, which extends a state's duties to any situation where actions under its control may cause a human rights impact in circumstances where that impact is foreseeable. Under this approach to causation, if a state has acted in a way which results in an increased risk of future harm, or has failed to take reasonable steps necessary to reduce a foreseeable future risk, then that will demonstrate a sufficient causal link between the state's conduct and the likely future human rights impact.

Another factor which would make causation easier to establish would be interpreting an environmental right as a right to a good or healthy environment in its own right, rather than as a right to an environment that facilitates the enjoyment of other human rights. As the *Demanda Generaciones Futuras v Minambiente* decision shows, if rights are given a more ecocentric framing, all that is required is to show that states have caused the environmental harm – it is not necessary to demonstrate a causal link to a particular human impact.

Interpreting state duties in a way that is informed by due diligence and the precautionary principle shifts the focus on to the procedural and preventative steps which states must take, rather than the ultimate consequences of their actions, thereby removing much of the challenge of causation. As long as there is adequate evidence to show a likely risk of future impacts on environmental rights, it is not necessary to prove a long-term causal chain.

Balancing Competing Rights

Any attempt to take care of future generations' rights will inevitably raise questions about how to prioritise their interests in relation to those of current generations. In some situations, actions which benefit future generations will also benefit current generations, and vice versa. For example, regulating the use of hazardous chemicals to prevent contamination of soil and water is obviously good for the health of people now and in the future. But in other cases (perhaps even more often than not) current generations may need to pay some price or forego some benefit in order to protect the interests of future generations. The climate crisis provides ample illustrations of this. We must invest in the transition to renewable energy now in order to ensure a safe climate for future generations, and the longer we delay that transition the greater the cost will be – both in terms of the severity of climate impacts and the monetary and social cost of dealing with them. The resources needed to take climate action could instead be used to address other human rights issues, something which is clearly most acute for poorer countries and communities who have significant, possibly urgent human rights demands. This of course points to the need for a global, cooperative effort to address climate change, so that poorer countries can be supported to take action without compromising their progress across other human rights issues. But it also demonstrates that, in order to include future generations within the framework of IHRL, some principled approach to balancing competing interests is required.

With the exception of the small category of rights which are considered to be peremptory norms (such as protections from torture, slavery and genocide), most human rights can be legally restricted when doing so is necessary to protect or advance other human rights, provided the restrictions are proportionate. Determining proportionality is especially challenging when one side of the balancing act involves the interests of future people, given the uncertainty that surrounds their circumstances, capabilities, resources and vulnerabilities. In economic contexts, future costs and benefits have typically been discounted compared to current costs and benefits. This is done through the application of a social discount rate, which aims to give a present-day value to future costs and benefits by factoring in the effect of interest, inflation, opportunity costs, and other economic variables.

However, the application of a social discount rate in the context of long-term environmental harm, particularly the impacts of greenhouse gas emissions, has received growing criticism.[72] Discounting the costs which will accrue to future generations ignores the scientific reality that climate change will continue to increase in severity and will be exacerbated the longer we delay taking stronger action. The prospect that our actions might lead to the transgression of planetary boundaries threatens irreversible, potentially existential damage that cannot be quantified. Furthermore, attempting to translate social impacts into economic terms ignores the fact that future people cannot participate in decisions which will affect them.[73] The rights and interests of future generations shouldn't be discounted in comparison to those of current generations, as doing so would be inconsistent with human rights principles of equality and universality. If we accept that future generations have human rights, then there is no morally significant reason to treat those rights any differently from those of present generations.[74]

72 Nicholas Stern, *The Economics of Climate Change: The Stern Review* (Cambridge University Press, 2007); Simon Caney, 'Human Rights, Climate Change and Discounting' (2008) 17(4) *Environmental Politics* 536; Moritz A Drupp et al, 'Discounting Disentangled' (2018) 10(4) *American Economic Journal: Economic Policy* 109.

73 See discussion in *Waratah Coal Pty Ltd v Youth Verdict Ltd and Others (No 6)* (n 29) 1216.

74 Simon Caney, 'Human Rights, Responsibilities and Climate Change' in Charles R Beitz and Robert E Goodin (eds), *Global Basic Rights* (Oxford University Press, 2009) 227, 240.

It should not be assumed that the rights of current and future generations will always be incompatible, however. In many ways, the interests of future generations will be improved by actively improving the lives of current generations. Edith Brown Weiss recognised the need to assure intra- as well as inter-generational equity, noting the strong link between poverty and environmental degradation and the fact that addressing poverty and disadvantage now should lead to better lives for future people as well. But there will be some instances where action to protect future rights will come at some cost to current generations, and vice versa. In these situations, we can start by going back to general principles of human rights law which determine when a limitation of rights will be justified.

To begin with, an action which is likely to cause harm to future generations or restrict their human rights needs to be justified on the basis that it is necessary to achieve a legitimate aim – usually the need to respect, protect or fulfil other human rights. Governments should be able to point to a benefit for current generations that is based in human rights (rather than, for instance, something which is merely politically popular or discretionary). Next, the restriction on future generations' rights needs to be proportionate to that aim, and not create a disproportionate burden for future generations. Small gains for present generations should not be pursued where they would be likely to result in major losses for future generations.[75] In assessing proportionality, it is important that future generations' rights are not discounted, but are given equal consideration to those of current generations.

Some additional parameters can be applied around this proportionality approach, designed to safeguard the basic needs of both current and future generations. While it's not appropriate to prioritise the rights of current generations over those of future generations simply on temporal grounds, it may be appropriate to prioritise basic needs where these are threatened. Henry Shue talked about this in terms of prioritising 'subsistence rights' over 'luxury rights'.[76] Under this approach, current generations might need to accept some lowering of their standard of living where that is necessary to protect the fundamental needs of future people.

A couple of other principles can be added to this approach. Firstly, the duty to respect and protect human rights should be viewed as a minimum standard of obligation, in some circumstances allowing for the duty to fulfil human rights to be postponed. This would ensure a base level of protection for rights – especially basic rights to subsistence and an adequate standard of

[75] United Nations Secretary-General (n 51) 13.

[76] Henry Shue, 'Subsistence Emissions and Luxury Emissions' (1993) 15(1) *Law and Policy* 39.

living – before pursuing more advanced levels of enjoyment, and not at the expense of the fundamental needs of future generations.

A risk with this approach is that governments may seek to defer action on costly rights – such as health, education and social security – on the basis that resources need to be conserved for the benefit of future generations. But this would only be justified where it could be shown that deferring current rights fulfilment was in fact necessary. Where possible, states should take action that achieves mutual benefits – it should only be in very specific and limited circumstances that a conflict between current and future generations' rights truly exists. In many cases there will be a way forward that benefits both current and future generations.[77] Lawrence has proposed an approach along these lines – what he calls the 'structural reform' approach – under which states are encouraged to implement long-term systemic reform while maintaining minimum standards of protection for current generations.[78] Where it is not possible to find a way to benefit both current and future generations, then priority should be given to safeguarding fundamental needs through the duty to respect and protect.

Together, the concepts of due diligence and the precautionary principle provide additional guidance in striking a better balance between the rights of current and future generations. Drawing on these principles, if a risk assessment for a proposed act or decision identifies that there is a risk of future harm to environmental rights, preventative and precautionary measures need to be taken, even if the act or decision would produce benefits for current generations.

Chapter 4 identified that one of the barriers to protecting future generations' environmental rights is that human rights courts and tribunals tend to give states wide discretion to determine their own environmental priorities. This is most obvious in the European doctrine of the margin of appreciation.[79] When it comes to determining an appropriate environmental policy, the European Court of Human Rights will be unwilling to substitute its own assessment for

[77] United Nations Secretary-General (n 51) 13, 18.

[78] Peter Lawrence, *Justice for Future Generations: Climate Change and International Law* (Edward Elgar, 2015) 78.

[79] *Handyside v the United Kingdom* [1976] (Application No 5493/72, 7 December 1976); *Hatton and Others v the United Kingdom* [2003] Eur Court HR, Application No 36022/97; Dean Spielmann, 'Allowing the Right Margin: The European Court of Human Rights and the National Margin of Appreciation Doctrine: Waiver or Subsidiarity of European Review?' (2012) 14 *Cambridge Yearbook of European Legal Studies* 381; Andrew Legg, 'Different Approaches to Deference in International Human Rights Law' in *The Margin of Appreciation in International Human Rights Law* (Oxford University Press, 2012) 38–66.

that of the state itself.[80] Instead, it grants states a 'margin of appreciation' to determine how to strike a fair balance between environmental protection and other interests and, provided they comply with their own domestic environmental laws, will largely endorse states' judgment in this regard.[81]

Even beyond Europe, international bodies give deference to states' priorities about resource allocation and other policies. This is necessary in order to respect states' legitimate and democratically made decisions, as an extension of respect for state sovereignty. But these domestic decisions can be influenced by the effects of 'short-termism', creating issues for the protection of future generations' rights. The combination of recommendations outlined earlier ought to help encourage more long-term thinking in the way these policies and priorities are determined. The balancing principle in particular allows for states to determine the best way to meet the needs of competing human rights. In the same way that the margin of appreciation requires states to strike a fair balance while still maintaining minimum standards, the theory set out earlier will embed core obligations which deliver intergenerational equity. Applying this approach, states should be viewed to have violated their obligations to future generations when their actions will foreseeably limit the enjoyment of human rights in a way which isn't proportionate to a human rights benefit for current generations.

5.5 CONCLUSION

This chapter has proposed a number of reforms, additions and interpretations for IHRL to support the protection of future generations' environmental rights. These begin with a recognition that future generations have rights and the introduction of a pathway for representative claims on behalf of future generations. I have suggested that future generations must be recognised as possessing the full range of environmental rights, and that these can be interpreted through a lens of intergenerational responsibility informed by the work of Edith Brown Weiss and the Maastricht Principles.

Under this approach, the obligations to respect, protect and fulfil the environmental rights of future generations would require that states undertake due diligence to assess the likely risk to future generations' environmental

80 Ole W Pedersen, *European Court of Human Rights and Environmental Rights* (Edward Elgar, 2019) 468.

81 *Dubetska v Ukraine* [2011] Eur Court HR 30499/03, 141; Hana Müllerová, 'Environment Playing Short-Handed: Margin of Appreciation in Environmental Jurisprudence of the European Court of Human Rights' (2015) 24(1) *Review of European, Comparative & International Environmental Law* 83.

rights. Where there is a credible risk of harm, states should take necessary measures to prevent that harm or mitigate its seriousness, provided those measures are reasonably available. Where serious and/or irreversible harm is likely and no preventative or mitigation measures can reasonably be taken, states should avoid the conduct altogether.

When there is tension between the rights of current and future generations, future generations' rights must not be discounted and must be treated on an equal basis to those of current generations. Fundamental needs may take priority over more advanced fulfilment of rights, provided it can be shown that limiting the latter is necessary to safeguard the former. Wherever rights are limited in order to protect other rights, those measures must be proportionate to the aim being pursued. Applying these principles, it should be possible to determine if a state has violated its obligations to respect, protect and fulfil the environmental rights of future generations. The case studies examined in Chapters 6, 7 and 8 will demonstrate how this can be applied in practice.

6

Unconventional Natural Resource Extraction

A Case Study of the Future Human Rights Impacts of Fracking and Deep Seabed Mining

6.1 INTRODUCTION

Chapter 5 presented a proposal for revising international human rights law (IHRL) in order to better protect the environmental rights of future generations. In the following chapters, this approach will be road-tested through a series of case studies. This first case study explores unconventional natural resource extraction, specifically deep seabed mining and hydraulic fracturing (fracking). As a case study, unconventional resource extraction like fracking and deep seabed mining allows us to consider the role of the precautionary principle and how environmental rights ought to deal with risk and uncertainty. It also examines how environmental rights apply to environmental impacts which might not have a direct impact on individuals' lives or livelihoods. For instance, would the loss of a species which is endemic to the hydrothermal vents of a particular deep-sea area be considered an interference with environmental rights? What rights do future generations have with respect to groundwater located beneath remote areas currently untouched by modern development? When we still have so much to learn about the interconnectedness of the Earth's various ecosystems, how should we be making decisions about activities that might cause irreparable harm? This case study will attempt to answer some of these questions through the application of the theory developed in Chapter 5.

6.2 HYDRAULIC FRACTURING (FRACKING)

6.2.1 *What Is Fracking?*

Hydraulic fracturing, also known as fracking, is a technique involving the injection of water and other substances at high pressure into rock to create fissures from which natural gases can be extracted. It is commonly used to extract gases (mostly methane) from shale deposits deep underground, but can also be used for coal-bed methane, coal seam gas and 'tight oil'. Shale gases are unconventional because the shale rock is much harder than other substances which contain gas (such as sandstone or limestone), and the cavities which contain the gas are typically much smaller and less connected, making it much more difficult to extract large volumes of gas.[1]

To extract the gas, drilling pads are first constructed by levelling off the ground above shale deposits. Wells are drilled into the ground, and there can be several wells drilled per pad. After drilling down to depths of up to 3,600 metres, drilling turns 90° and proceeds horizontally through the shale deposits. Large volumes of fluid are pumped at high pressure into these horizontal cavities to fracture the rock and allow gas to flow. Fracking fluids consist of water with the addition of various chemicals which help reduce friction and facilitate high pressures. The fluid also contains 'proppants', which are small particles (usually fine silica sand) that help to keep the fissures open and allow the gas to flow more freely. Hydraulic fracturing can occur for several days and can be repeated multiple times for each well.[2] The gas flows back up the bore holes into collection wells where it is then transported through pipelines for processing and distribution.[3]

After being injected through the horizontal well, around one third of the fracking fluid returns to the surface, usually over a period of two to four weeks. This fluid is known as 'flowback', and it can retain high levels of the chemicals and proppant used in the initial fracking. As well as the flowback, wells can

1 Grantham Research Institute on Climate Change and the Environment, 'What Is Shale Gas, How Is It Extracted through Fracking and What Are Fracking's Impacts?' (11 February 2022) <www.lse.ac.uk/granthaminstitute/explainers/what-is-shale-gas-how-is-it-extracted-through-fracking-and-what-are-frackings-impacts>; Benjamin K Sovacool, 'Cornucopia or Curse? Reviewing the Costs and Benefits of Shale Gas Hydraulic Fracturing (Fracking)' (2014) 37 *Renewable and Sustainable Energy Reviews* 249, 250–251; Matt Ridley, *The Shale Gas Shock* (Global Warming Policy Foundation, 2011) <www.thegwpf.org/images/stories/gwpf-reports/Shale-Gas_4_May_11.pdf>.

2 Ridley (n 1); Sovacool (n 1).

3 British Geological Survey, 'Shale Gas Extraction', *British Geological Survey* <www.bgs.ac.uk/geology-projects/shale-gas/shale-gas-extraction>; Ridley (n 1) 7.

also bring to the surface 'produced water', which is naturally occurring groundwater that rises up with the gas through the well. The flowback and produced water can bring with them various substances from beneath the earth's surface, including toxic and radioactive materials.[4] Wastewater is stored in tanks or pits at the well site before either being recycled for further fracking or transferred to treatment facilities. Treated wastewater is often disposed of into sewerage systems.[5] During production, the gas can be flared (a controlled burn that is employed for testing and safety purposes) or vented deliberately into the air. There can also be leakage of gas from wells and pipelines.[6]

Fracking projects require a network of wells and pipelines across a large area to extract the gas from large basins. After collection at the well site, the gas is piped to compressor stations which process the gas from multiple wells and move it on to larger pipelines, which often stretch vast distances to get the gas to processing and export facilities. The construction and operation of the well sites involve considerable road infrastructure to facilitate movement of trucks and machinery. Depending on the available water on site, trucks may need to bring in large volumes of water for fracking and may be used to remove wastewater for treatment and disposal. Consequently, drill sites can resemble heavy industrial zones with heavy traffic and associated noise and emissions.[7]

While the process of unconventional gas extraction involves a series of techniques across multiple phases, hydraulic fracturing is a critical component. Although there are different uses of the term, Short et al explain that 'fracking' has come to represent more than just the specific technique of hydraulic fracturing and now is commonly used to refer to the full process.[8] Proponents of gas developments regularly use the narrower definition, however, referring only to the specific technique of hydraulic fracturing, while opponents often use it to refer to the wider network of wells, roads and infrastructure as well as the activities and impacts associated with fracking projects.

For the purposes of this chapter, fracking will use the definition adopted by the Permanent People's Tribunal on Fracking (discussed in more detail later): 'unconventional extraction of shale gas, coal-bed methane, coal seam gas and

4 Damien Short et al, 'Extreme Energy, "Fracking" and Human Rights: A New Field for Human Rights Impact Assessments?' (2015) 19(6) *The International Journal of Human Rights* 697, 707.

5 Ridley (n 1) 21–22; Sovacool (n 1) 251, 262.

6 Melissa Denchak, 'Fracking 101', *NRDC* (19 April 2019) <www.nrdc.org/stories/fracking-101>.

7 Sovacool (n 1) 255; Richard A Kerr, 'Natural Gas from Shale Bursts onto the Scene' (2010) 328(5986) *Science* 1624.

8 Short et al (n 4) 702–703.

"tight oil"' – extending beyond the specific technique to include the broader operation of extraction.[9]

6.2.2 *Where Is Fracking Occurring? What Is the Uptake?*

The use of fracking has grown considerably in the last twenty years, with significant projects in the United States, Canada and Australia where it is used to extract gas for domestic electricity generation and export. These countries possess large basins with rich shale gas deposits. In the United States, the Marcellus Shale covers a vast area underneath much of New York, Pennsylvania, Ohio, Maryland and West Virginia, stretching into neighbouring states as well. The Utica basin sits beneath Ohio, New York, Pennsylvania and West Virginia, and extends into Kentucky, Tennessee, Virginia and Maryland and as far north as Ontario and Quebec. There are also basins in North Dakota and Texas with rich shale gas deposits. In Canada, there are significant deposits in the Montney and Duverney basins in Alberta, and fracking has been used in Canada since the 1950s. Some fracking has occurred in the United Kingdom, mostly in northern regions, particularly Lancashire, but there are also sizeable deposits in the south and in Scotland. China and Argentina also produce gas through fracking.

In Australia, fracking is occurring in the Beetaloo Basin in the Northern Territory. There are also large shale gas deposits within the Surat Basin (located in Queensland and New South Wales), the Cooper Basin (Queensland and South Australia) and Georgina Basin (Queensland and Northern Territory), with a number of exploration permits in place in these regions. These basins overlap with the Great Artesian Basin, the world's largest underground aquifer, spanning 1.7 million square kilometres. The Canning Basin in Western Australia also has large shale gas deposits.

Because of the environmental and health impacts detailed later, and because there remains much uncertainty about the long-term impacts of fracking, several jurisdictions have banned fracking or implemented moratoria.[10] For example, the New York state legislature imposed a full permanent ban in 2021,

9 Thomas A Kerns and Kathleen Dean Moore (eds), *Bearing Witness: The Human Rights Case against Fracking and Climate Change*, 1st ed. (Oregon State University Press, 2021) 21–22.

10 Hector Herrera, 'The Legal Status of Fracking Worldwide: An Environmental Law and Human Rights Perspective', *Global Network for the Study of Human Rights and the Environment* (6 January 2020) <https://gnhre.org/human-rights/the-legal-status-of-fracking-worldwide-an-environmental-law-and-human-rights-perspective>; Hector Herrera, '¿En qué va el fracking en el mundo?', *Razón Pública* (22 July 2019) <https://razonpublica.com/en-que-va-el-fracking-en-el-mundo>.

following the announcement of a ban by the Governor in 2014.[11] Bans are also in place in Vermont, Maryland, Oregon and Washington. The Delaware Basin Commission, membership of which includes the governors of Delaware, New Jersey, New York and Pennsylvania as well as a representative of the Federal Government, banned fracking in the Delaware Basin in 2021. The following year, it also banned the disposal of fracking wastewater within the watershed of the Delaware River.[12] The Canadian provinces of Nova Scotia and Quebec have banned fracking. Bans are also in place in many European countries, including Bulgaria, France, Denmark, the Netherlands, Germany and Ireland.[13]

In the Northern Territory in Australia, home to several significant sites of natural and cultural heritage, including Kakadu National Park, a moratorium on fracking was introduced in 2016 following considerable public opposition to proposed developments. A scientific inquiry was commissioned to look at the risks of fracking and the resulting report (published in 2018) recommended 135 measures which would need to be implemented in order to manage the risks of fracking.[14] In response, the NT Government accepted the Inquiry's recommendations and lifted the moratorium.

The Australian Senate held an inquiry into fracking in the Beetaloo Basin, which recommended that stronger environmental controls were needed.[15] At the time of writing, the Northern Territory has not yet implemented all of the recommendations from these various inquiries, but has approved the expansion of fracking in the Beetaloo Basin.[16] Other Australian states,

[11] Marisa Guerrero, 'New York State Codifies Fracking Ban in Budget', *NRDC* (3 April 2020) <www.nrdc.org/experts/marisa-guerrero/new-york-state-codifies-fracking-ban-budget>; Thomas Kaplan, 'Citing Health Risks, Cuomo Bans Fracking in New York State', *The New York Times* (17 December 2014) <www.nytimes.com/2014/12/18/nyregion/cuomo-to-ban-fracking-in-new-york-state-citing-health-risks.html>.

[12] Sahana Rao, 'Delaware River Commission Further Guards Basin from Fracking', *Natural Resources Defence Council* (19 December 2022) <www.nrdc.org/bio/sahana-rao/delaware-river-commission-further-guards-basin-fracking>; Herrera, 'The Legal Status of Fracking Worldwide' (n 10).

[13] Herrera, 'The Legal Status of Fracking Worldwide' (n 10).

[14] Scientific Inquiry into Hydraulic Fracturing in the Northern Territory, *Final Report* (Northern Territory Government, 2018) ('*Scientific Inquiry into Hydraulic Fracturing in the Northern Territory*').

[15] Senate Standing Committees on Environment and Communications, 'Oil and Gas Exploration and Production in the Beetaloo Basin'; Australian Senate Standing Committees on Environment and Communication, *Oil and Gas Exploration and Production in the Beetaloo Basin: Second Progress Report* (April 2022).

[16] Lisa Cox, 'NT Government Deal to Buy Beetaloo Basin Gas from US Company Labelled "Carbon Bomb"', *Guardian Australia* (online, 23 April 2024) <www.theguardian.com/australia-news/2024/apr/23/nt-government-beetaloo-basin-gas-sale-plan-tamboran-resources>;

Victoria and Tasmania, have banned fracking, while New South Wales imposes very strict restrictions. The Queensland Government recently announced a plan to ban fracking in the fragile channel country of the Kati Thanda–Lake Eyre Basin, but still supports fracking in other regions.[17]

In January 2015, the Scottish Government put in place a moratorium on unconventional oil and gas development in Scotland, following the publication of a report by an independent, expert, scientific panel.[18] In 2019 the United Kingdom Government transferred all licensing power for oil and gas exploration in Wales to the Welsh Assembly, which has since decided not to grant licences for fracking, essentially creating a ban on the practice.[19]

Elsewhere in the UK, a moratorium on fracking has been in place across England since 2019, largely because of scientific uncertainty about the risk of fracking-induced earthquakes. Short-term Conservative Prime Minister Liz Truss had briefly lifted the ban on fracking in 2022, claiming it was necessary to address energy shortages.[20] The lifting of the ban prompted strong opposition, including from within the Conservative Party, and it was reinstated by her successor Rishi Sunak.[21] The British Labour Party has been strongly opposed to fracking, and after the change of government in 2024 it is likely the ban will remain in place.

While there are these examples of bans and moratoria, enthusiasm for fracking remains in many places, often driven by the desire of the fossil fuel industry to extract as much as they can while greenhouse gas policies still allow for profits to be made and costs passed on to consumers. It is also driven by the diminishing supply of conventional resources, which are much easier

Roxanne Fitzgerald, 'US Gas Company Tamboran Receives Approval to Expand Fracking in Beetaloo Basin', *ABC News* (online, 7 June 2024) <www.abc.net.au/news/2024-06-07/nt-government-approves-tamboran-exploratory-fracking-expansion/103950262>.

17 'Increased Protections, Greater Certainty for Globally Iconic Queensland Section of Lake Eyre Basin', *Ministerial Media Statements* <https://statements.qld.gov.au/statements/99391>; RT Kingsford and AJD Walburn, *Oil and Gas Exploration and Development in the Lake Eyre Basin: Distribution and Consequences for Rivers and Wetlands, Including the Coongie Lakes Ramsar Site* (Research Paper, CSIRO, 4 October 2022) <https://doi.org/10.1071/MF22063>.

18 The Scottish Government, *Independent Expert Scientific Panel – Report on Unconventional Oil and Gas* (The Scottish Government, 2014).

19 Welsh Government, 'Licensing Powers on "Fracking" Transferred to Wales' (online, 1 October 2018) <https://gov.wales/licensing-powers-fracking-transferred-wales>.

20 Aubrey Allegretti and Sandra Laville, 'Liz Truss to Lift Fracking Ban "Despite Little Progress on Earthquake Risk"', *The Guardian* (online, 16 September 2022) <www.theguardian.com/environment/2022/sep/15/liz-truss-to-lift-fracking-ban-despite-little-progress-on-earthquake-risk>.

21 'What Is Fracking and Why Is It Controversial?' *BBC News* (online, 26 October 2022) <www.bbc.com/news/uk-14432401>.

to extract, and by disruptions to energy supply caused by the Russian war against Ukraine (among other geopolitical threats).

However, the window for this practice to remain profitable is narrowing, which appears to have accelerated efforts by the fossil fuel industry to 'get in while they can'.[22] It is a somewhat perverse consequence of the global push to reduce greenhouse gas emissions that this aspect of fossil fuel extraction has in fact accelerated in recent years. Given the royalties that flow to governments from resource developments, it is not surprising that some have also sought to 'fast track' unconventional gas extraction to maximise revenue. In Australia, the then Morrison Government touted a plan for a 'gas-led recovery' following the COVID-19 pandemic, a plan which has hardly been dampened by the new Albanese Government. With these forces apparently encouraging more rapid roll-out of fracking projects, there is an increased risk that the short- and long-term risks may not be given due consideration. Section 6.2.3 will outline the nature of some of those risks.

6.2.3 *What Are the Impacts of Fracking?*

The moratoria and bans summarised above point to the considerable uncertainty and concern that surrounds the environmental, social and economic risks of fracking. The long-term impacts of fracking are uncertain, but studies have uncovered a range of impacts for individuals, families and communities affected by fracking. Many stories of people's experiences living near gas fields were shared through witness testimony at the Permanent Peoples' Tribunal on Human Rights, Fracking and Climate Change.[23] The Permanent People's Tribunal (PPT) is a body of experts assembled to hear human rights claims and is independent of any state's legal system.[24] It has been used to litigate cases where conventional legal action is not available. While it cannot give legal decisions or remedies, it can comment on the legality of state policies or actions and serves as a powerful means of recording evidence and documenting people's experiences.

The PPT on Fracking was convened in 2018 and asked to give an expert opinion on the impacts of fracking on human rights. The Tribunal heard from witnesses from around the world, who shared stories of contaminated water, depleted groundwater reserves, disease and death of livestock, increased rates

[22] Short et al (n 4).

[23] Kerns and Moore (n 9); *The Court's Judgments* (Permanent Peoples' Tribunal on Human Rights, Fracking and Climate Change, Permanent Peoples' Tribunal).

[24] 'Permanent Peoples' Tribunal' <https://permanentpeoplestribunal.org/?lang=en>.

of childhood disease, cancers and respiratory diseases, and struggles to seek justice. The stories shared with the PPT are supported by scientific studies and peer-reviewed research.[25]

The effects of fracking on communities can be understood as infringing on the enjoyment of environmental rights – both the right to a healthy environment and the environmental dimensions of other recognised rights. There are also implications for civil and political rights, including participation in democratic processes, freedom of expression and the right to protest. Some of these impacts are relatively short term, but others have the potential to persist and impinge on the rights of future generations. Where the long-term risks are less certain, this raises the question of how to safeguard future generations' rights in the face of scientific uncertainty.

Impacts on Water

Water impacts are a major concern in areas where fracking occurs, many of which are susceptible to drought and water insecurity. Fracking requires huge volumes of water, which come either from nearby surface or underground sources or are trucked in from elsewhere. Where local sources are used, this can lead to competition for water between fracking operations and household or agricultural uses, which can exacerbate existing water insecurity.[26] If water for fracking is taken from underground aquifers, depletion can have long-term consequences, since the recharge rates of these aquifers are often very slow, particularly in arid regions.[27] Water is essential to the enjoyment of a range of human rights. It's fundamental for both the right to health and the right to an adequate standard of living.[28] Under the Convention on the Rights of the Child, states are obliged to ensure access to safe drinking water for all children, as part of the right to health.[29] The potential for fracking to interfere with the right to water has been recognised by the United Nations Special

[25] For a detailed discussion of the PPT testimony, see Kerns and Moore (n 9).

[26] Sovacool (n 1) 257.

[27] Jose Rodriguez, Joonghyeok Heo and Kee Han Kim, 'The Impact of Hydraulic Fracturing on Groundwater Quality in the Permian Basin, West Texas, USA' (2020) 12(3) *Water* 796; Short et al (n 4) 706.

[28] *International Covenant on Economic, Social and Cultural Rights* 1966 arts 11 and 12 ('ICESCR'); Committee on Economic Social and Cultural Rights, *General Comment No. 15: The Right to Water* (2003) [8]; Catarina de Albuquerque, *Report of the Special Rapporteur on the Human Right to Safe Drinking Water and Sanitation, Catarina de Albuquerque. Addendum: Mission to the United States* (UN, 2 August 2011) <https://digitallibrary.un.org/record/710677>.

[29] *Convention on the Rights of the Child* 1989 art 24 ('CRC').

Rapporteur on the Right to Water.[30] Clearly, where water supplies are depleted the potential for human rights interference is high.

In addition to concerns about depletion, water contamination is a significant issue. In a report issued by the United States Environmental Protection Agency (EPA) in 2016, evidence of water contamination was identified

> in all stages of the process: water withdrawals for hydraulic fracturing; spills during the management of hydraulic fracturing fluids and chemicals; injection of hydraulic fracturing fluids directly into groundwater resources; discharge of inadequately treated hydraulic fracturing wastewater to surface water resources; and disposal or storage of hydraulic fracturing wastewater in unlined pits, resulting in contamination of groundwater resources.[31]

As the EPA report indicates, contamination can occur either through underground transference into groundwater or through overland flow. The chemicals that are used in fracking fluids contain many potentially harmful substances which could leach into groundwater. In some jurisdictions, mining companies have been shielded from having to disclose the nature and concentration of chemicals used in their fracking fluids. Reporting requirements have improved in recent years, but some lack of transparency remains as to the precise make-up of chemicals used in the fracking process.[32]

The studies which have been done have shown the presence of dangerous hydrocarbons and heavy metals.[33] These chemicals have been linked to a range of serious health impacts, including damage to the brain and nervous systems, impacts on reproductive systems, cancers, sensory organ damage, gastrointestinal and liver disease, and skin irritation.[34] As Finkel and Law have

[30] de Albuquerque (n 28).

[31] Patrick G Lee, 'EPA Concludes Fracking a Threat to U.S. Water Supplies', *ProPublica* (online, 14 December 2016) <www.propublica.org/article/epa-concludes-fracking-a-threat-to-u.s.-water-supplies>.

[32] Christopher B Hill, Om P Yadav and Eakalak Khan, 'Hydraulic Fracturing Chemical Disclosure Policy and Data Analysis: Metrics and Trends in Transparency' (2021) 55(6) *Environmental Science & Technology* 3918; Alexis L Maule et al, 'Disclosure of Hydraulic Fracturing Fluid Chemical Additives: Analysis of Regulations' (2013) 23(1) *New Solutions* 167.

[33] Madelon L Finkel and Adam Law, 'The Rush to Drill for Natural Gas: A Public Health Cautionary Tale' (2011) 101(5) *American Journal of Public Health* 784; Madelon L Finkel and Adam Law, 'The Rush to Drill for Natural Gas: A Five-Year Update' (2016) 106(10) *American Journal of Public Health* 1728.

[34] Roxana Witter et al, *Potential Exposure-Related Human Health Effects of Oil and Gas Development: A White Paper* (Colorado School of Public Health, 1 January 2008); Finkel and Law, 'The Rush to Drill for Natural Gas: A Public Health Cautionary Tale' (n 33).

explained, some of these problems can take months or even years to materialise and, where chemicals disrupt the endocrine system, they can affect development and cause other harm to children not yet born.[35] Exposure can occur through drinking, cooking or bathing, as well as through watering of livestock and horticultural use. There is also a risk that the methane being extracted through fracking will seep into aquifers used for water supplies, creating the risk of explosion when it reaches a certain concentration and exsolves out of the water.[36]

When the fracking fluid returns as flowback it can contain many of these chemicals. Further, the produced water that comes to the surface after fracking occurs can bring with it toxic or radioactive substances, or have high salinity. Management of these waste waters must be carefully done to avoid contamination of surface waters. Problems can arise at storage sites or in transit where leaks can easily lead to contamination of lakes, rivers, creeks or wetlands. The potential for contaminated water to enter the landscape is exacerbated by climate change, given the increased frequency and severity of rains and flooding. In some instances, wastewater has been dumped directly into streams or onto roads, where the potential for contaminating drinking water supplies or environmental water is high.[37]

As noted earlier, the Northern Territory Government in Australia responded to community concerns about fracking by convening a scientific inquiry to gather more insight into the risks of fracking. The inquiry concluded that there was unequivocal evidence that above-ground leakages of chemicals would occur, either through human error, failure of infrastructure or extreme weather. While the report concluded that the risk of such contamination impacting on drinking water or ecological wetlands/rivers was low, it noted that this depended on the distances and other design features which needed to be assured in order to keep risks to a minimum.[38]

Wastewater management and the construction of fracking infrastructure like roads and well pads can also impact on water flows that have important ecological value. In Australia, serious concerns have been raised about the potential impact of proposed fracking developments in the Channel Country of the Lake Eyre Basin – a vast network of desert rivers and ephemeral

35 Finkel and Law, 'The Rush to Drill for Natural Gas: A Public Health Cautionary Tale' (n 33).

36 United States Environmental Protection Agency, *Hydraulic Fracturing for Oil and Gas: Impacts from the Hydraulic Fracturing Water Cycle on Drinking Water Resources in the United States (Final Report)* (Reports and Assessments, December 2016) 26 <https://cfpub.epa.gov/ncea/hfstudy/recordisplay.cfm?deid=332990>.

37 Sovacool (n 1) 257.

38 Scientific Inquiry into Hydraulic Fracturing in the Northern Territory (n 14).

wetlands that drain into Kati Thanda (also known as Lake Eyre) following monsoonal rains. The potential for fracking activities to disrupt this seasonal water flow or contaminate the rivers, lakes and wetlands is of serious ecological concern.[39] In response to these concerns, the Queensland Government announced a plan to ban unconventional gas exploration in the Kati Thanda–Lake Eyre Basin.[40]

Where contamination or depletion of natural water occurs, it can impact on the cultural rights of Indigenous peoples. First Nations people from the Northern Territory in Australia have shared stories of how fracking has impacted on their traditional fishing and hunting practices, sacred sites, songlines and other cultural practices. Speaking in relation to fracking in the NT's Beetaloo Basin, Traditional Owner Johnny Wilson has said: 'We are so concerned for our Country, our water, our sacred sites, and our cultural heritage, which we know will be damaged if fracking goes ahead'.[41] In Queensland, the Kati Thanda–Lake Eyre Basin covers the traditional lands of many Aboriginal groups. Traditional Owner from the Mithaka People, Josh Gorringe, said: 'We don't want to see fracking on the floodplains, because it will impact the waterbird dreaming. Fracking will impact this songline. This is our culture, and you can't put a price on it'.[42]

Air Pollution

The various phases of fracking operations create different problems in terms of air pollution and associated health risks. Drilling and extraction activities produce a range of pollutants, including hydrogen sulfide, sulfur dioxide and volatile organic compounds (VOCs), which can be harmful. The fracking process releases BTEX (benzene, toluene, ethylbenzene and xylene) – a group of VOCs found in fossil fuels that can be harmful to human health, including through causing various cancers, birth defects and disorders of the blood or

39 Kingsford and Walburn (n 17); Michael Slezak and Marty Smiley, 'Once a Decade Lake Eyre Comes Alive. But Scientists Say the Gas Industry's Final Push Is Putting It All at Risk', *ABC News* (online, 3 October 2022) <www.abc.net.au/news/2022-10-04/lake-eyre-impact-of-gas-development/101486164>.

40 'Increased Protections, Greater Certainty for Globally Iconic Queensland Section of Lake Eyre Basin' (n 17).

41 Samantha Dick, 'Gas Industry CEO Tells Senate Inquiry NT Fracking Cannot Happen without Traditional Owners' Consent', *ABC News* (online, 22 March 2022) <www.abc.net.au/news/2022-03-23/nt-beetaloo-basin-senate-inquiry-gas-industry-ceo/100929398>.

42 Lock the Gate, 'Lake Eyre Basin Protections Must Be Priority for Soon to Be Premier Steven Miles' (12 December 2023) <www.lockthegate.org.au/lake_eyre_basin_protections_must_be_priority_for_soon_to_be_premier_steven_miles>.

nervous system. These chemicals can potentially flow with the gas or leak from compressor stations.[43] BTEX is also present in diesel exhaust, and can be a product of diesel-powered pumps and vehicles used in fracking operations. The fumes from these engines also produce ground-level ozone, the main component of smog, which causes asthma and other respiratory conditions.[44] Exposure to air pollution caused by fracking can lead to adverse birth outcomes, including preterm birth and low birthweight, especially in women living close to gas wells.[45]

There is also potential for the extracted gas to contain radon, a radioactive gas which can make its way into households through the on-sale of the produced gas.[46] Radon is found naturally underground but can be brought up through fracking. It is a known cause of lung cancer.[47]

Severe respiratory conditions can also be caused through exposure to fine crystalline silica, which occurs during the mining of sand for use as a proppant in the fracking fluid.[48] Inhaling silica can lead to silicosis, lung cancer, tuberculosis and other chronic pulmonary diseases, and serious cases can be fatal.

Methane venting and leakage and emissions from flaring contribute to climate change. Methane is a greenhouse gas with a high potential for global warming. It is more potent than CO_2 in the short term, though CO_2 persists longer in the atmosphere. It is estimated to have a global warming potential around twenty-five times greater than CO_2 over a 100-year period, and even more than this in the shorter term.[49] The end use of natural gas (mostly for energy generation) also contributes to global warming, as burning produces CO_2. The process of fracking thereby contributes to climate change both in the short and long term. The widespread human rights impacts of climate change are well documented.[50]

43 Kerns and Moore (n 9) 25–29.
44 Sovacool (n 1) 257; Short et al (n 4) 708 709.
45 Finkel and Law, 'The Rush to Drill for Natural Gas: A Five-Year Update' (n 33) 1729; Short et al (n 4) 713.
46 Sovacool (n 1) 258.
47 Kerns and Moore (n 9) 29.
48 Finkel and Law, 'The Rush to Drill for Natural Gas: A Five-Year Update' (n 33) 1729.
49 Ramón A Alvarez et al, 'Greater Focus Needed on Methane Leakage from Natural Gas Infrastructure' (2012) 109(17) *Proceedings of the National Academy of Sciences* 6435.
50 John Knox, 'Climate Change and Human Rights (Annual Report to Human Rights Council)'; David R Boyd, *A Safe Climate: Human Rights and Climate Change. Report of the UN Special Rapporteur on Human Rights and the Environment* (United Nations General Assembly, 15 July 2019) <www.srenvironment.org/report/a-safe-climate-human-rights-and-climate-change>.

Impacts on Land and Soil

The toxic and/or radioactive substances that pose a threat through contamination of water also present risks if they are found in soil used for agriculture or surrounding residential communities. The process of fracking can bring to the surface potentially harmful substances which normally remain underground, or rapidly accelerate the natural processes through which they might normally rise up.[51] Fracking has also been linked to increased seismic activity, with potential for fracking to trigger earthquakes up to 50 kilometres away.[52]

Rights to land are also a significant issue that frequently arises in relation to fracking. In most jurisdictions, ownership of minerals beneath the land's surface belongs to the state. Holders of other forms of title have limited rights to oppose or benefit from the extraction of gas beneath the ground. The struggles of farmers and pastoralists to resist gas developments on their land are well documented across many countries.[53] While they might receive monetary compensation if wells or other infrastructure are constructed within the boundaries of their properties, this is often not available if they merely live in proximity to a gas development (in spite of the fact that their air and water may be affected, as detailed earlier).

Indigenous peoples' rights to their lands can be negatively affected in various ways. Under the United Nations Declaration on the Rights of Indigenous Peoples (UNDRIP), Indigenous peoples have recognised rights to control and develop their traditional lands and resources and to have these conserved.[54] States are obliged to protect Indigenous peoples' relationship with the land through appropriate methods of legal recognition.[55] However,

[51] Tom Myers, 'Potential Contaminant Pathways from Hydraulically Fractured Shale to Aquifers' (2012) 50(6) *Groundwater* 872.

[52] Gillian Foulger, 'Fracking Can Cause Earthquakes Tens of Kilometres Away – New Research', *The Conversation* <http://theconversation.com/fracking-can-cause-earthquakes-tens-of-kilometres-away-new-research-116539>; Anna Holligan, 'Groningen Gas Fields – the Dutch Earthquake Zone', *BBC News* (online, 19 May 2013) <www.bbc.com/news/world-europe-22542982>; Pathikrit Bhattacharya and Robert C Viesca, 'Fluid-Induced Aseismic Fault Slip Outpaces Pore-Fluid Migration' (2019) 364(6439) *Science* 464.

[53] Michelle Maloney, 'Contaminated Life: The True Cost and Human Rights Impacts of Unconventional Gas Mining in Australia', *Center for Humans and Nature* (20 April 2018) <https://humansandnature.org/contaminated-life-the-true-cost-and-human-rights-impacts-of-unconventional-gas-mining-in-australia>; Maddelin McCosker and Victoria Pengilly, 'Origin Energy Backs Away from Outback Fracking Plans but Farmers Say Fight Isn't Over', *ABC News* (online, 22 September 2022) <www.abc.net.au/news/2022-09-23/origin-energy-backs-out-gas-exploration-channel-country-fracking/101456466>.

[54] *United Nations Declaration on the Rights of Indigenous Peoples* 2007 arts 26 and 29 ('*UNDRIP*').

[55] Ibid art 29.

UNDRIP does not grant a right to Indigenous peoples to benefit from mineral extraction that occurs on their land, other than as can be negotiated through the requirement for free, prior and informed consent (FPIC).[56] FPIC is required before any hazardous materials can be stored or disposed of on Indigenous lands,[57] but it is generally understood that the UNDRIP does not grant a veto power to Indigenous peoples with respect to proposed extractive developments.[58] And studies around the world show that states frequently have not complied with the obligation contained in article 32 of UNDRIP with respect to consulting and cooperating in good faith in order to obtain Indigenous peoples' consent.[59] As a result, fracking operations frequently take place through processes and in locations which are incompatible with Indigenous land tenure.

Furthermore, having regard to the significant environmental impacts of fracking outlined earlier, unconventional gas extraction has potential to significantly disrupt Indigenous peoples' cultural rights with respect to their lands and waters. This includes the right of Indigenous peoples to transmit their cultural and spiritual traditions to future generations, which is recognised in article 27 of the ICCPR as well as in the UNDRIP.[60] These rights are also threatened by global warming, as recognised by the Human Rights Committee in its decision relating to Torres Strait Islanders in 2022.[61]

Governance and Procedural Rights

Other human rights concerns arise relating to the governance of fracking. Within both human rights and environmental law, a number of procedural rights are in place to help to ensure that decision-making is transparent and decision-makers accountable, and through this process to try to minimise

56 Ibid art 32.

57 Ibid art 29.

58 S James Anaya, 'Report of the Special Rapporteur on the Situation of Human Rights and Fundamental Freedoms of Indigenous Peoples'; Mauro Barelli, 'Free, Prior and Informed Consent in the Aftermath of the UN Declaration on the Rights of Indigenous Peoples: Developments and Challenges Ahead' (SSRN Scholarly Paper No 1991731, 25 January 2012) <https://papers.ssrn.com/abstract=1991731>; Daria Shapovalova, 'Indigenous Rights and Resource Extraction: One Step Forward Two Steps Back', *Environmental Policy and Law* (7 December 2020) <https://environmentalpolicyandlaw.com/news-blog/indigenous-rights-and-resource-extraction-one-step-forward-two-steps-back>.

59 Stephen Young, *Indigenous Peoples, Consent and Rights: Troubling Subjects* (Routledge, 2020).

60 *UNDRIP* (n 54) arts 11, 13 and 25.

61 *Daniel Billy and Others v Australia* [2022] Human Rights Committee Communication No 3624/2019.

negative impacts on human rights. These environmental procedural rights are usually described as a trio of rights comprising access to information, participation and access to justice.[62]

Access to information – International environmental law recognises access to information as a key procedural right attached to environmental decision-making.[63] Accurate information about likely environmental impacts is necessary to make well-informed decisions, and people must have access to this information to be able to participate in and challenge decisions which are likely to adversely affect the environment. One of the challenges revealed through the testimony before the PPT has been the inadequacy of impact assessments which fail to consider the full range of environmental, social and cultural impacts of fracking. As discussed later, there still remains considerable uncertainty about the nature and scale of risks associated with fracking, and even where assessments have been done these are not always made publicly available.

A key concern raised by witnesses at the PPT was secrecy around the operations and impacts of fracking developments. For a long time, particularly in the United States, extraction companies were not required to disclose the techniques or chemicals used in fracking, on the basis that these were trade secrets entitled to protection.[64] This had practical ramifications: witnesses at the PPT shared stories of leaks and explosions that were ineffectively dealt with because responders had to wait until they could obtain information about the chemicals they were dealing with.[65] Such secrecy also meant that communities could not accurately assess the risk of living nearby to fracking developments and could not contribute fully in debates and decision-making.[66] More broadly, lack of transparency in decision-making can easily undermine trust in those processes and the parties involved.[67] While disclosure requirements

62 *Convention on Access to Information, Public Participation in Decision-Making and Access to Justice in Environmental Matters* 1998 ('*Aarhus Convention*').

63 Ibid.

64 Finkel and Law, 'The Rush to Drill for Natural Gas: A Public Health Cautionary Tale' (n 33); Finkel and Law, 'The Rush to Drill for Natural Gas: A Five-Year Update' (n 33); Maule et al (n 32); Dusty Horwitt and Meilen Teklemichael, *Chemical Makers' Exemptions from Fracking Chemical Disclosure Rules* (Physicians for Social Responsibility, July 2022) <https://psr.org/wp-content/uploads/2022/07/chemical-makers-exemptions-from-fracking-chemical-disclosure-rules.pdf>.

65 Kerns and Moore (n 9) 74–75.

66 Simona Perry, 'Self-Determination and the Right to Information on the Shale Gas Frontier', *Center for Humans and Nature* (3 March 2017).

67 Miriam R Aczel and Karen E Makuch, 'Human Rights and Fracking in England' (2018) 20(2) *Health and Human Rights* 31, 34.

have improved in many jurisdictions, companies still commonly withhold information about the full nature of chemicals used.[68]

There also remain issues about monitoring the impacts of fracking underground and across ecosystems, and ensuring that these studies are publicly available. Reliable and regular monitoring of environmental impacts (including robust baseline studies prior to approval or commencement of development) is an essential part of fulfilling the right to access to information.

Right to participation – A key procedural right in environmental matters is the right to participate in decision-making. As noted earlier with respect to Indigenous peoples, this right is usually translated into the right to free, prior and informed consent. The link to access to information is clear: for consent to be free and informed there must be adequate and accurate information upon which to judge the likely risks and benefits. The right to participate extends to any person, household or community who is likely to be adversely affected by the decision, as well as to the general public. It entails an obligation on the state to notify publicly when an application for a proposed development has been submitted and to facilitate adequate opportunities for the public to contribute to the decision-making process.[69]

Importantly, the *Aarhus Convention on Access to Information, Public Participation in Decision-Making and Access to Justice in Environmental Matters* specifies that '[e]ach Party shall ensure that in the decision due account is taken of the outcome of the public participation'.[70] This means that participation goes beyond simply hearing the concerns of the public, and must include a genuine consideration of those concerns in decision-making. In the testimony to the PPT, many witnesses told stories of their struggles to have their concerns about fracking heard. They described difficulty obtaining information about planned developments and the lack of opportunities to communicate their views. A number of witnesses pointed to the apparent influence of mining companies on local decision-making and the power imbalance between large, wealthy corporations and small communities without financial resources to fight back.

Some witnesses even gave evidence of actions that had been taken by mining companies and local authorities to intimidate them into not speaking out or to retaliate against them when they did.[71] Indigenous people in the

68 Kevin Trickey, Nicholas Hadjimichael and Prachi Sanghavi, 'Public Reporting of Hydraulic Fracturing Chemicals in the USA, 2011–18: A before and after Comparison of Reporting Formats' (2020) 4(5) *The Lancet Planetary Health* e178.

69 *Aarhus Convention* (n 62) art 6.

70 Ibid art 6(8).

71 Kerns and Moore (n 9) 119–132.

United States have given accounts of the state violence they have been subjected to when standing up for their land rights. For example, Wenonah Hauter has told the story of the Standing Rock Sioux who opposed fracking on their lands and were subjected to tear gas, concussion grenades, attack dogs and water cannons in freezing temperatures.[72]

A related issue is that of increasing anti-protest laws, which have been imposed in several jurisdictions and applied for various forms of environmental protest, including against fracking.[73] Such laws not only violate procedural rights relating to participation, but also impinge upon substantive human rights like the right to freedom of expression and the right to peaceful assembly.[74] They have been commonly enforced against people attempting to halt coal and gas developments, often in cases where other channels for participation and legal challenge have proved fruitless. In the context of climate change, some of these protests have been successfully defended on the basis that they are justified by the urgent need to address the climate emergency.[75] But in general, anti-protest laws operate to stifle open consultation and deliberation and increase the power imbalance between mining operators and those affected by their operations.

One of the health impacts that can be easily overlooked is the effect that fracking can have on the mental health of people living near gas developments, particularly those who have fought against fracking in and near their communities. Testimony presented to the PPT session on fracking showed clearly the chronic stress and anxiety that people have suffered living near to fracking operations, stemming often from their deep concern for the long-term environmental impacts and their sense of loss as they see ecosystems and

[72] Wenonah Hauter, 'Fracking's Inhumanity', *Center for Humans and Nature* (3 March 2017) <https://humansandnature.org/frackings-inhumanity>.

[73] Bob Brown, 'Harsher Anti-Protest Laws Targeting Environmentalists Are Putting Greed before Green', *The Guardian* (online, 27 June 2022) <www.theguardian.com/commentisfree/2022/jun/27/harsher-anti-protest-laws-targeting-environmentalists-are-putting-greed-before-green-bob-brown>; Hauter (n 72); Short et al (n 4) 214.

[74] *International Covenant on Civil and Political Rights* 1966 arts 19 and 21 ('*ICCPR*').

[75] Sandra Laville, 'Extinction Rebellion Founder Cleared over King's College Protest', *The Guardian* (online, 9 May 2019) <www.theguardian.com/environment/2019/may/09/extinction-rebellion-founder-cleared-over-kings-college-protest>; John Vidal and environment editor, 'Not Guilty: The Greenpeace Activists Who Used Climate Change as a Legal Defence', *The Guardian* (online, 10 September 2008) <www.theguardian.com/environment/2008/sep/11/activists.kingsnorthclimatecamp> ('Not Guilty'); Nicole Rogers, 'Activists Are Using the Climate Emergency as a New Legal Defence to Justify Law-Breaking', *The Conversation* <http://theconversation.com/activists-are-using-the-climate-emergency-as-a-new-legal-defence-to-justify-law-breaking-122949>.

livelihoods affected. In Australia, the campaign against fracking became so distressing for some landholders that it even led to suicide.[76]

Fracking can also have a significant impact on the well-being of Indigenous peoples, whose special connection with Country and obligations to care for that Country can be dramatically affected by the environmental damage caused by fracking. These concerns have an intergenerational dimension, since Indigenous worldviews often involve a sense of obligation towards future generations and even a non-linear conception of time.[77] These impacts on mental well-being are exacerbated where affected people are not given opportunities to participate in decision-making and do not feel that their concerns have been listened to.

Access to justice – The third dimension of environmental procedural rights is access to justice, which requires that people negatively affected by an environmental decision have recourse to a remedy. This has also proved a challenge in relation to fracking, often linked to the weak property and participation rights that are in place. Once fracking has commenced and environmental and health impacts have been observed, it has proved difficult for affected individuals and communities to seek redress, noting that in some cases damage is irreversible or very difficult to restore.

Strong procedural rights are crucial to ensuring meaningful protection of other human rights. They help to promote transparency and accountability in decision-making and combat corruption and secrecy. They are difficult to implement for the benefit of future generations, however, since they are unable to participate in discussions and are reliant on current generations to represent their interests. Section 6.2.4 considers what the issues are for future generations in relation to fracking and how we ought best to address them.

6.2.4 *Litigating the Future Human Rights Impacts of Fracking*

The impacts of fracking described earlier occur across a broad timeframe. Some can be felt immediately, while others take years or even decades to materialise. Some impacts can be rectified, while others are irremediable. Direct exposure to chemicals can cause immediate health impacts like asthma attacks or skin irritation. Longer-term exposure could lead to more chronic conditions that develop over time, including cancers and brain or lung

[76] Maloney (n 53).

[77] Mary Graham, 'Some Thoughts about the Philosophical Underpinnings of Aboriginal Worldviews – AHR' [2008] (45) *Australian Humanities Review* 2 <http://australianhumanitiesreview.org/2008/11/01/some-thoughts-about-the-philosophical-underpinnings-of-aboriginal-worldviews>.

conditions. Such prolonged exposure could also cause developmental delays and birth defects, leading to intergenerational consequences. Clean-up of sites may not be possible where chemicals have leached into groundwater or flowed into river and wetland systems. And of course, the contribution of unconventional gas extraction to climate change has very long-term consequences. Yet for all the impacts of fracking that have been studied, there remains some uncertainty about its long-term effects. It is not known precisely how long-term exposure to fracking chemicals might impact on people's health or the environmental consequences which could result from disruptions to the water table. However, there is sufficient evidence to suggest that future generations may face significant negative human rights impacts as a consequence of fracking. This section imagines human rights litigation that seeks to address the impacts of fracking on future generations, using the framework developed in Chapter 5.

As a starting point, under the system of IHRL imagined in Chapter 5, the law would recognise that future generations will possess all of the same environmental rights as current generations. These include rights to life, health, food, water, culture and self-determination, among others. These rights would all be understood to exist in a mutually reinforcing relationship with the environment, such that environmental degradation can constitute a violation of rights where relevant obligations have been breached.

Future generations would fall under the jurisdiction (and therefore the obligations) of the state which has control over fracking activities. This would be established through the three limbs of jurisdiction outlined in Chapter 5. First, fracking could be shown to be under the effective control of the state, either because it is carried out by state-owned entities or based on the regulatory power the state exercises over private actors. Secondly, as outlined earlier, there is sufficient evidence to demonstrate that fracking will affect the rights of future generations. Third, that evidence should also be adequate to demonstrate that the impact on future generations' rights is foreseeable to the state. Consequently, this revised approach to jurisdiction will mean that the state owes duties with regard to the rights of future generations.

In terms of the specific nature of those duties, we can get some more detail by unpacking the tripartite framework of obligations, informed by intergenerational equity, due diligence and the precautionary principle. As explained in Chapter 5, the duties to respect, protect and fulfil can be informed by Edith Brown Weiss's concept of intergenerational equity, which entails equity of options, quality and access. The duty to respect involves an obligation to refrain from conduct that will limit the diversity or quality of the environment that future generations will enjoy or their ability to access environmental goods. The duty to protect requires the state to regulate the actions of private entities in order to preserve diversity, quality and access, while the duty to fulfil imposes a positive obligation to maintain, rehabilitate and improve environmental diversity, quality and access.

With respect to fracking in particular, these obligations are implicated by the likelihood that fracking will deplete the water table and pollute the soil, water and air. Fracking is also likely to damage ecosystems and traditional lands. In most cases, these impacts occur within the bounds of state regulations – that is, they are permitted consequences of approved activities. It is therefore strongly arguable that by permitting fracking, the state has prima facie failed to meet its obligations to respect, protect and fulfil the rights of future generations, as they are understood through a lens of intergenerational responsibility.

To demonstrate that the state has breached its obligations, it will be necessary to go a little deeper into the specifics of what it was required to do and how it has failed to perform its duties. This is where standards of due diligence and the precautionary principle are useful. Due diligence requires that the state conducts an adequate assessment of both environmental and human rights impacts. Under a more future-focused approach, the human rights assessment needs to consider the long-term impacts of fracking, including its contribution to climate change. It should also consider the cumulative impacts of different projects and any other relevant contextual factors.

Once the risk to human rights has been identified, the state needs to take appropriate steps to prevent or minimise that impact. If there is a risk of severe or irreparable damage (which is arguably the case with all fracking), then the activities shouldn't proceed. Where there is doubt about the long-term effects, the precautionary principle dictates that precautionary steps still need to be taken. In relation to fracking, the risk of harm comes from the process of hydraulic fracturing itself, as well as from the way that gas and waste products are stored and transported. While some precautions could be taken in the management of gas wells and their outputs, it is arguable that fracking by its very nature is hazardous to human rights and cannot be carried out safely.

In terms of causation, it will be enough to show that the state has permitted fracking in circumstances where future human rights impacts were foreseeable. The incorporation of principles of intergenerational equity, due diligence and precaution means that it is not necessary to show that the harm has materialised or is imminent. The relevant duties here are preventative and forward-looking – states must act to conserve environmental goods for the benefit of future generations. To use the language of due diligence, these are duties of conduct, not of result. A breach will be caused where the state acts in a way which does not adequately guard against foreseeable harm. For impacts associated with climate change, it is not necessary to show that the state's actions are the sole or direct cause of those impacts. Instead, causation could be established by showing that the state made a contribution to climate change which it knew or should have foreseen.

In its defence, a state might argue that the economic benefits that flow from the extraction of gas through fracking justify the future human rights impacts. For such an argument to succeed, the state would need to demonstrate that fracking was necessary to advance economic rights. It is questionable whether the current market conditions surrounding fossil fuels bear this argument out, especially if there are other industries or projects that could yield similar economic gains. Furthermore, potential economic benefits need to be assessed against the economic costs of climate change and of the other environmental damage caused by fracking. As explained in Chapter 5, the fact that a project might yield short-term economic benefits needs to be balanced against the long-term costs, and those long-term costs should not be discounted on a pure time basis. This would obviously need to be assessed on a case-by-case basis, and human rights bodies will typically give considerable weight to a state's own judgment about these things, but once we look across the longer time scale to weigh up any economic benefits of fracking against the negative human rights consequences, it is difficult to conclude that the impacts are justified. As such, it is likely that a state which permits fracking will be in breach of its duties to future generations.

6.3 DEEP SEABED MINING

The second example of unconventional resource extraction considered here is deep seabed mining. The deep seabed is located beyond the continental shelf and can be found at depths of between 4,000 and 6,000 metres below sea level. The trenches and canyons located in these areas can be 10,000 metres deep or more. Mining of the deep seabed is intended to extract minerals which are increasingly scarce in terrestrial deposits. These minerals are particularly in demand for use in technology manufacture, including renewable energy technology like solar panels and batteries.

6.3.1 *What Is Deep Seabed Mining?*

In the deepest areas of the ocean, the seabed contains abundant mineral deposits. Many of these are found in potato-sized nodules comprising multiple substances (polymetallic nodules) such as manganese, nickel, copper and cobalt. These nodules can also contain lithium, molybdenum and other rare-earth elements. Another source of metal deposits on the seabed are massive sulphide deposits that form at the boundaries of tectonic plates and other active areas like volcanic islands. They are the product of hydrothermal vents, where superheated seawater emerges from beneath the oceanic crust, having absorbed metal sulphides from the surrounding rock. After being expelled from the vents, the deposits settle on the sea floor. They contain high concentrations of metals including zinc, gold, silver, lead, arsenic, copper and cobalt.[78] A third source of metals are polymetallic crusts. These form on the slopes of seamounts (underwater mountains that rise from the sea floor) and can be several inches thick. They can contain significant amounts of manganese, iron, copper, cobalt, nickel, platinum and other metals and rare earth minerals.[79]

While these deposits have been known about for decades, it is only in recent years that advances in technology have made their extraction feasible, particularly in the context of diminishing terrestrial reserves and increased market demand for critical minerals. That being said, much of the technology and processes that would be used are proprietary and difficult to learn about except in quite general terms. For nodules, mining would likely be done through a remotely controlled collector (or 'crawler') that is driven along the sea floor to pick up the nodules and then transport them to the surface.[80] Polymetallic sulphides have been recovered using an excavator on the seafloor which can separate and crush the deposits. These are then pumped in a slurry to the surface.[81] Mining of polymetallic crusts is more complicated, since the crusts would require considerable force to remove them from the hard substrates they have attached to over millennia. The process for separating the crust would likely be quite environmentally destructive.[82]

6.3.2 *How Is Deep Seabed Mining Regulated?*

The deep seabed has been characterised as the common heritage of mankind in the *United Nations Convention on the Law of the Sea* (UNCLOS). At the same time, UNCLOS recognises the rights of states to explore and exploit the

78 Luc Cuyvers et al, *Deep Seabed Mining* (International Union for the Conservation of Nature, 2018) 7–10.

79 Ibid 9; Pradeep A Singh, 'Deep Seabed Mining and Sustainable Development Goal 14' in Walter Leal Filho et al (eds), *Life below Water* (Springer International Publishing, 2022) 271.

80 Cuyvers et al (n 78) 23–24.

81 Ibid 25–26.

82 Ibid 24.

mineral deposits that lie within it. The legal rules which regulate activities in the deep seabed depend on where those deposits are located in relation to states' maritime boundaries. Because it is located off the continental shelf, most of the deep seabed is beyond the limits of national jurisdiction, in what is called the Area.[83] To govern the Area, UNCLOS established the International Seabed Authority (ISA) and gave it the task of developing regulations for deep seabed mining. To date, the ISA has been unable to settle on regulations, though the ISA Council has said it is working with a view to finalising them in 2025.[84] The ISA also has power to issue exploration licences for the deep seabed in the Area, and has adopted some regulations to govern exploration activities. Without finalised exploitation regulations, no deep-sea mining projects are operational in the Area, but a number of private and state-owned companies have licences to conduct exploration.[85]

Where a state has only a narrow continental shelf, like some Pacific Island nations, New Zealand and Japan, some of the deep seabed may occur within the exclusive economic zone (EEZ). If this is the case, the state would have rights to exploit the resources within its EEZ, including through issuing licences to private corporations.[86] UNCLOS places some limits on this, requiring that the national regulations be at least as strict as the standards set by the ISA, and that they meet the overarching obligation in UNCLOS to protect and preserve the marine environment.[87] UNCLOS requires that all states minimise pollution and protect and preserve fragile ecosystems and the habitat of threatened or endangered species.[88] Some pilot projects have been attempted by states that have deep-sea deposits within their EEZs.

6.3.3 *What Are the Risks of Deep Seabed Mining?*

There is much uncertainty about the environmental impacts of deep seabed mining, largely due to the significant gaps in our understanding about deep-sea ecology and biodiversity. In particular, there is still much to be learned about what lies beneath the seabed and in the deepest canyons and trenches,

[83] *United Nations Convention on the Law of the Sea* 1982 1 ('UNCLOS').

[84] International Seabed Authority, 'Draft Regulations on Exploitation of Mineral Resources in the Area: Consolidated Text' <www.isa.org.jm/wp-content/uploads/2024/02/Consolidated_text.pdf>.

[85] Pradeep A Singh, 'The Invocation of the "Two-Year Rule" at the International Seabed Authority: Legal Consequences and Implications' (2022) 37(3) *The International Journal of Marine and Coastal Law* 375.

[86] Cuyvers et al (n 78) 33.

[87] *UNCLOS* (n 83) 192.

[88] Ibid 194.

and how these species might contribute to ecosystem services.[89] Van Dover has explained that repeated surveys of a single site regularly uncover new species, and many species appear in very small numbers within a bulk sample. As a result, there is a 'very large proportion of rare, poorly known and typically undescribed species, even at vents that have been studied for several decades'.[90] A recent study of the biodiversity of the Clarion–Clipperton Zone (a massive area of deep seabed that is subject to multiple exploration licences) found that of the more than 5,000 species recorded there, at least 85 per cent of them were previously unknown to science.[91]

It is unclear exactly how deep-sea species may be affected by the disturbance of the sea floor or increased levels of sediment, noise and vibration, but it is expected that some environmental degradation is unavoidable.[92] Deep-sea environments can be home to very specialised species, with unique variations isolated to particular locations. The special nature of deep-sea species and ecosystems makes it extremely difficult to locate suitable offset sites, and the locations make rehabilitation challenging. Studies to date have indicated that species recovery in deep-sea areas can be very slow, and in some cases a species may never recover. This creates a significant risk that deep seabed mining will lead to extinctions.[93] As Cuyvers et al have argued, 'the interconnectivity of ocean biodiversity leads to the assumption that the extinction of a species, or the destruction of an ecosystem, could have severe repercussions'.[94] Because there is still so much to learn about these deep-sea ecosystems and the unique species that live within them, another risk of mining is that we will lose the opportunity to learn about what is currently unknown.[95]

It is also not clear how far some of the effects of deep seabed mining may spread and what impact they might have on coastal areas or fishing zones. For example, turbidity plumes from disturbed sediment may travel considerable

89 Cuyvers et al (n 78) 31.

90 Cindy Lee Van Dover, 'Mining Seafloor Massive Sulphides and Biodiversity. What Is at Risk?' (2011) 68(2) *ICES Journal of Marine Science* 341, 342–343.

91 Muriel Rabone et al, 'How Many Metazoan Species Live in the World's Largest Mineral Exploration Region?' (2023) 33(12) *Current Biology* 2383.

92 Cuyvers et al (n 78) 41.

93 CL Van Dover et al, 'Scientific Rationale and International Obligations for Protection of Active Hydrothermal Vent Ecosystems from Deep-Sea Mining' (2018) 90 *Marine Policy* 20; Van Dover (n 90); Holly J Niner et al, 'Deep-Sea Mining with No Net Loss of Biodiversity – An Impossible Aim' (2018) 5 *Frontiers in Marine Science* 1; Ann Vanreusel et al, 'Threatened by Mining, Polymetallic Nodules Are Required to Preserve Abyssal Epifauna' (2016) 6(1) *Scientific Reports* 26808.

94 Cuyvers et al (n 78) 41.

95 Van Dover (n 90) 342.

distances and last for long periods of time, with potentially widespread effects on marine ecosystems of all kinds.[96] Because of these risks, many groups and nations have expressed concerns about going ahead with deep seabed mining. This includes some small island communities, who are worried about how disturbance of the deep seabed might impact on their surrounding marine environments. For example, the Kioa Climate Emergency Declaration, drafted by a coalition of Pacific leaders, calls on governments to (among other things) adopt 'ocean policies that are compatible with the climate goals including banning deep-sea mining and other destructive practices that endanger the survival of small island communities'.[97]

6.3.4 *Litigating the Future Human Rights Impacts of Deep Seabed Mining*

In terms of the environmental rights of future generations, deep seabed mining raises a different set of issues from fracking. First, there is much more uncertainty about the environmental impacts of deep seabed mining. This is due to a combination of our less advanced understanding of deep-sea ecology and geology and our more limited evidence base for the activity itself. Like fracking, though, it has potential to cause serious, potentially irreversible environmental harm, so litigation is likely to be used to prevent it from occurring rather than to seek a remedy after the fact.

Secondly, many of the minerals that are targeted by deep seabed mining are needed for technology that will play an important role in the global transition to renewable energy. Given the profound human rights impacts of climate change, there is arguably a human rights-based justification for progressing decarbonisation. Deep seabed mining could therefore be justified on the grounds that it helps avert more serious human rights consequences of severe climate change.

A third factor is that, because deep seabed mining is planned primarily for uninhabited areas beyond national jurisdiction, its impacts on human rights might be more diffuse. Whereas fracking impacts directly on the water, air and soil that people rely on for their basic needs, deep seabed mining creates environmental consequences without such direct, visible human rights effects. Conceptualising deep seabed mining as a human rights violation

96 Cuyvers et al (n 78) 64.

97 'Kioa Climate Emergency Declaration' <www.ohchr.org/sites/default/files/documents/issues/climatechange/cfi-hrc-53-session/submissions/2022-11-28/PIANGO.pdf>; 'Pacific Island Countries Make Historic Grassroots Declaration on Climate Change Emergency in Kioa, Fiji' 6 <www.caritas.org.au/news/media-releases/pacific-island-countries-make-historic-grassroots-declaration-on-climate-change-emergency-in-kioa-fiji>.

depends in large part on how we define environmental rights. In the absence of a broadly framed right to environment, it might be challenging to show that deep seabed mining impacts on human rights, especially across a longer timeframe.

As a case study, deep seabed mining points to the importance of an ecocentric framing for environmental rights. As discussed in Chapter 3, the right to a healthy environment is not formally recognised in IHRL, and the soft law articulations of the right are quite anthropocentric in their framing. A more ecocentric interpretation of the right to a healthy environment would likely be of more use for a problem like deep seabed mining, which impacts on areas of common concern, beyond national jurisdictions, and with potentially system-wide consequences. Deep seabed mining also highlights the role of due diligence and the precautionary principle in managing uncertainty and risk. Even where the activities occur within national boundaries, the impacts may be felt in other countries, raising issues of both inter-temporal and extraterritorial obligations. Whichever rights and duties we rely on to try to litigate deep seabed mining, such litigation will require a somewhat novel approach.

As a starting point, we need to consider who the likely respondent would be in any litigation. This could be based on the state with effective control over the activities. If the exploration or extraction is planned within national boundaries, the respondent would be that state, on the basis that it exercises control over the mining activities through the granting of licences and otherwise regulating those activities. If the activity is in the Area, then the appropriate respondent is likely to be the state that has jurisdiction over the mining companies involved, on the basis that they can exert some control over the activities of their corporate nationals.

The victims may not be nationals of those states, however. The impacts of deep seabed mining are more likely to affect island nations or those with coastlines on the affected oceans. For the case to be admissible, claimants would need to show that the prospective respondent state owes duties to people in those countries. Such extraterritorial duties are easy enough to demonstrate through the approach adopted by the Committee on the Rights of the Child in the *Sacchi* complaint, discussed in Chapter 5. This involves a three-step test of showing (1) effective control over activities, (2) a negative human rights impact caused by those activities, and (3) foreseeability of the negative impact. As was outlined in Chapter 5 and discussed earlier in relation to fracking, a variation on this approach can also be used to establish inter-temporal jurisdiction which provides a basis for obligations towards future generations. For the purposes of standing, future generations of another country could be victims of anticipated deep seabed mining if it could be argued that they will likely suffer a human rights impact caused by the mining which was foreseeable by the state concerned.

A key issue will therefore be the foreseeability of impacts on human rights and how to frame an argument when we don't know with any certainty what the impacts will be, because the activity in question is largely untested. However, while we don't know for sure how impacts will be felt, the scientific understanding we do have suggests that increased sediment in the water from disruption of the deep seabed could foreseeably impact on marine and coastal ecosystems, even large distances away. If this occurs, it is foreseeable that it would impact on rights to food, health, culture and even the right to life, of people who rely on those ecosystems. Furthermore, the fact that we don't have certainty around these impacts should not be interpreted as a lack of foreseeability. As noted in Chapter 5, the precautionary principle should influence the way we approach foreseeability of harm as well: a lack of scientific certainty should trigger a more, not less, cautious approach.

The issue of foreseeability also makes it clear how valuable a standalone right to a healthy environment would be in litigation like this. While the impacts on rights to food, health, and culture for instance, might be difficult to predict, we can be much more confident that deep seabed mining will be destructive of deep-sea habitat and the unique species that live there. If we take a narrow view of environmental rights which only encompasses the ways in which the environment underpins the enjoyment of other rights, then it could be hard to argue that the loss of niche, deep-sea ecosystems is a human rights issue. After all, could we really say we suffer a loss from the extinction of a species which we were not even aware of? Yet our increasing understanding of the interconnectedness of the biological world indicates that we may very well suffer from the loss of these unique species and ecosystems. We don't yet know what role they might play in complex systems. And the resources that are being extracted through deep seabed mining have taken millennia to build up as a result of specific, place-based processes. Once these are gone they can never be replaced and the deep-sea environment will be changed forever. With a human right to a healthy environment, these facts themselves would be enough to show a negative human rights impact. And because some impact on human rights is foreseeable, we could establish duties for the states concerned, even though the specific details may be unknown. From this point we can also argue that future generations should be considered 'victims' for the purposes of bringing a claim through appropriate representatives.

The content of duties with respect to future generations' rights is similar to those discussed in relation to fracking, which drew on due diligence and the precautionary principle. The state would be obliged to assess the risk and take appropriate steps to prevent any negative impact on human rights. Given the level of uncertainty and potential for severe and irreversible environmental harms, it is arguable that only a moratorium on deep seabed mining would be an appropriately precautionary approach.

A final issue to consider is any possible justification based on the contribution of deep-sea minerals to decarbonisation of energy supplies and mitigation of climate change. A state might argue that the human rights consequences are justified because they are a necessary and proportionate consequence of efforts to protect human rights from the impacts of climate change. This is a very difficult claim to assess. For one thing, the countries likely to be most affected by deep seabed mining are also likely to endure the worst impacts of climate change. For another, weighing up opposing factors is difficult when there is uncertainty on both sides of the equation.

Significant weight should be given to the potential for extraction to impact entire ecosystems and cause the loss of whole species. This is a serious impact on the right to a healthy environment. Furthermore, we still have so much to learn about the ecology of the deep seabed that its degradation would not only be a loss for knowledge's sake alone, but could also mean a loss of potential benefits when we consider the role that deep-sea species might play in wider systems and their possible uses.

A human rights case relating to deep seabed mining would not be easy and may not succeed. It would need to confront a range of issues that are new for international human rights law. The assessment of whether mining activities are justified would ultimately depend on the specifics of each case, and that assessment may change as our scientific understanding changes. Much would also depend on the state of the law surrounding the right to a healthy environment, which is currently not yet recognised as a legally enforceable right. But thinking through these issues helps to articulate some of the arguments that could be made, if not in a legal tribunal or court then at least in advocacy and activism around deep seabed mining.

6.4 CONCLUSION

This chapter has considered the role of environmental rights in addressing the future impacts of unconventional natural resource exploitation, specifically fracking and deep seabed mining. As the case studies have shown, a revised approach to human rights would allow for state duties towards future generations to be articulated. States would be required to take adequate due diligence measures to minimise disproportionate impacts on the rights of future generations and, where those impacts could not be prevented or sufficiently mitigated, the activities could be considered a breach of human rights. Deep seabed mining in particular demonstrates the importance of incorporating the precautionary principle into IHRL, since that principle helps to navigate the challenges of decision-making in a context of considerable scientific uncertainty. It also shows that, where environmental harm is generally anticipated but the specific manifestations of that harm are unknown, an ecocentric right to a healthy environment would be a valuable tool for rights-based litigation. Both these examples have had connections to climate mitigation – fracking as an activity which contributes to climate change and deep seabed mining as a potentially necessary part of decarbonising our energy supply. Chapter 7 will build on this dimension by examining the ways in which we might litigate other aspects of climate change.

7

Climate Adaptation, Decarbonisation and Geoengineering

A Case Study to Explore Litigation Options for Climate-Related Human Rights Violations Affecting Future Generations

7.1 INTRODUCTION

The human rights impacts of climate change are diverse and, while some effects are already being felt, the worst impacts are yet to occur and will continue for generations to come, even if we succeed in cutting greenhouse gas emissions in the near future. As they currently stand, the nationally determined contributions (NDCs) submitted by nation states under the *Paris Agreement* are inadequate to meet the agreed objective of keeping global warming to 'well below 2°C'. The United Nations Environment Programme's latest Emissions Gap Report (released in late 2023) calculated that, if countries fully implemented their NDCs, global warming would at best be kept to 2.5°C above pre-industrial levels.[1] If all countries who have pledged to reach net zero succeed, global warming could be restricted to 1.8°C, but this is still above the preferred Paris target of 1.5°C. Moreover, widespread success on reaching net zero would require significant changes in countries that have set some very ambitious targets which they are currently not on track to achieve.[2]

[1] United Nations Environment Programme, *Emissions Gap Report 2023: Broken Record – Temperatures Hit New Highs, Yet World Fails to Cut Emissions (Again)* (20 November 2023) <www.unep.org/resources/emissions-gap-report-2023>; Katherine Calvin et al, *Climate Change 2023: Synthesis Report. Contribution of Working Groups I, II and III to the Sixth Assessment Report of the Intergovernmental Panel on Climate Change (Summary for Policymakers)* (Intergovernmental Panel on Climate Change (IPCC), 25 July 2023) 10.

[2] United Nations Environment Programme (n 1) ix.

There is therefore a very real risk that global warming will exceed safe levels, creating a potentially catastrophic situation for future generations. And, because of the prolonged presence of carbon in the atmosphere, even if greenhouse gas emissions were to stop immediately, future generations would still face the adverse impacts of past emissions for an extended period.

In some jurisdictions, environmental rights have successfully been used in legal cases to force governments to reduce greenhouse gas emissions and strengthen adaptation measures. This trend has begun to shift to the international level. However, to date, there has been only very limited success from international or regional rights-based climate litigation. This third and final case study will consider how an improved approach to international environmental rights might enable successful climate litigation and advocacy for future generations. It will include an analysis of climate litigation to date, focusing on cases brought by children and young people, and consider the extent to which these have been able to operationalise intergenerational justice in interpreting human rights obligations.

The case study will also consider how environmental rights might apply to our responses to climate change. As will be shown, decisions about how to implement the energy transition and what sorts of mitigation and adaptation measures to pursue have potentially significant impacts for future generations' rights. This is particularly the case for proposals to implement geoengineering technology. This case study will explore how the framework for intergenerational human rights responsibility articulated in Chapter 5 would apply to litigation in relation to climate mitigation and adaptation.

7.2 FUTURE ENVIRONMENTAL RIGHTS IMPACTS OF CLIMATE CHANGE

This section summarises the various ways that climate change will impact on the environmental rights of future generations. This includes not only the impacts of global warming itself, but also the potential impacts that could flow from our responses to it, including our efforts to decarbonise economies and adapt to a changing climate.

7.2.1 *Global Warming*

The future impacts of climate change are by now so thoroughly discussed in academic literature and mainstream discourse that it feels almost unnecessary to summarise them here. In the past decade the volume of research on the human rights implications of climate change has grown enormously, and it has been recognised by numerous United Nations bodies as the greatest human rights challenge of modern times. Rather than repeat the Intergovernmental Panel on Climate Change's (*IPCC*) forecasts about heatwaves, droughts, cyclones and sea levels, this section highlights a selection of issues that will face future generations in a changed climate. These issues bring implications for a wide range of environmental rights.

The environmental changes that are occurring and will continue into the future as a result of climate change threaten to create significant health risks for future generations. Global warming is likely to facilitate the spread of vector-borne diseases, including malaria, into previously unaffected regions. More extreme weather events will bring more rainfall and floods that may increase the prevalence of water-borne diseases such as cholera. Malnutrition and other illnesses are anticipated to impact child growth and development, while heat waves pose a serious health threat for older people.[3] The world which future generations will inherit is therefore likely to have new and more serious health problems which they will need to address.

Food and housing insecurity are also significant concerns. Losses of arable land due to changes in temperatures, rainfall and sea levels threaten crop production in Africa, Small Island States and other developing countries. Crops may also be affected by new pests or declining pollinator numbers. Fisheries are also expected to be affected by rising ocean temperatures and acidification.

Climate change is anticipated to exacerbate issues related to unsafe and inadequate housing. Certain communities and nations, including Arctic regions, Small Island States and low-lying mega-delta areas, are already highly susceptible to flooding, rising sea levels and storm surges, and these impacts are expected to intensify in the future. As rural livelihoods become increasingly unreliable, there will likely be additional pressure on housing as people migrate to urban areas. Currently, approximately 1 billion individuals reside in urban slums located in precarious areas, such as fragile hillsides or flood-prone riverbanks, making them more vulnerable to climate-related impacts. It is projected that climate change will force many more people to relocate to urban slums or informal settlements. These impacts on food and housing security will be felt worse by poor communities and developing states, adding another aspect of injustice to that facing future generations.

Climate change also threatens future generations' rights to culture, especially for Indigenous peoples. As discussed in Chapter 3, the right to culture entails the right to pass on cultural knowledge and practices to future

3 Calvin et al (n 1) 49–51.

generations. Where culture is connected to land, animals or ecosystems, a changing climate threatens the ability of future generations to learn from elders and perform their own cultural responsibilities.

Future generations will also face significant economic, political and social challenges in a world affected by climate change. The challenges outlined earlier are likely to affect poorer nations and communities more than wealthy ones, expanding the existing development gap. Resource scarcity may drive internal and international tensions, potentially leading to conflict. Extreme weather events will lead to more displacement of people and more migration. These problems will demand greater international assistance and cooperation, at the same time that domestic challenges are also more profound and pressing.

Climate change brings challenges for the enjoyment of a wide range of human rights, including rights to health, housing, food, water and culture, but the full impact on human rights will be more complex and extensive than a simple list like this can capture. Like all impacts of climate change, the human rights consequences for future generations will be compounding and cascading. They will experience climate-based interferences with their rights across multiple fronts, while also bearing the cost of rectifying past harm and attempting to slow the continuing progression of climate change.

7.2.2 *Decarbonisation*

In addition to the human rights which are affected by climate change itself, there are potentially significant human rights risks associated with our responses to climate change. Fully addressing the impact of climate change on the human rights of future generations requires that we consider the potential effects of the mitigation and adaptation measures which states might implement.

To begin, we can consider the potential human rights consequences of climate mitigation – the process of reducing emissions in an effort to prevent the worst effects of climate change. Mitigation is being pursued through the process of decarbonising economies. While this is an essential process in order to avoid catastrophic climate collapse, it brings with it a number of human rights risks that will need to be carefully navigated. And because most states have delayed decarbonisation, they are now facing an urgent need to transition quickly to a green economy, a context which presents its own range of human rights risks.

One way to consider the potential human rights impacts of mitigation is to look at the risks associated with decarbonisation across different sectors of the economy.[4] The renewable energy industry is an obvious place to start. Here, decisions about where to locate renewable energy infrastructure like solar arrays, hydroelectric dams and wind farms can present a number of human rights concerns, some of which have potentially long-term consequences. For example, a wind energy project in Norway was ruled unlawful because of its predicted impact on Indigenous peoples' rights to herd and hunt reindeer.[5] In Kenya, the legal basis for developing the Lake Turkana Wind Project was successfully challenged because of a failure to obtain the free, prior and informed consent of the local Indigenous peoples.[6] In Mexico, there have been concerns about the impact of wind farm developments on Indigenous people in the Oaxaca region for some time.[7] For example, an Indigenous community in Oaxaca, Unión Hidalgo, has commenced legal action in France against the French energy company, Électricité de France, arguing that it failed to obtain their free, prior and informed consent before commencing a project to develop a wind farm on their lands.[8] By contrast, the European Court of Human Rights has rejected claims that wind farms constitute a breach of the right to private and family life because the nuisance

4 For more detailed analysis, see Bridget Lewis, 'Human Rights' in Gerry Nagtzaam, Katie O'Bryan and Mark Beaufoy (eds), *Legal Pathways to Deep Decarbonisation in Australia* (LexisNexis, 2023).

5 *Statnett SF et al v Sør-Fosen sijte et al* [2021] Supreme Court of Norway HR-2021-1975-S.

6 *Kochale and Others v Lake Turkana Wind Power Ltd and Others* [2021] Environment and Land Court at Meru, Kenya Case No 163/2104.

7 Peace Brigades International, *Wind Farms and Concerns about Human Rights Violations in Oaxaca* (14 March 2014) <https://pbi-mexico.org/fileadmin/user_files/projects/mexico/files/PBI_Publications/1403BriefingWindFarmsPBI.pdf>; Stephanie Friede, *Enticed by the Wind: A Case Study in the Social and Historical Context of Wind Energy Development in Southern Mexico* (Mexico Institute) <www.wilsoncenter.org/publication/enticed-the-wind-case-study-the-social-and-historical-context-wind-energy-development>; European Centre for Constitutional and Human Rights, 'Wind Park in Mexico: French Firm Disregards Indigenous Rights' (2020) <www.ecchr.eu/en/case/wind-park-in-mexico-french-firm-disregards-indigenous-rights>; Business and Human Rights Resource Centre, *Renewable Energy and Human Rights Benchmark: Key Findings from the Wind and Solar Sectors* (2021) <https://media.business-humanrights.org/media/documents/2021_Renewable_Energy_Benchmark_v5.pdf>; Rosa Lehmann and Anne Tittor, 'Contested Renewable Energy Projects in Latin America: Bridging Frameworks of Justice to Understand "Triple Inequalities of Decarbonisation Policies"' [2021] *Journal of Environmental Policy & Planning* 1; Sofia Avila-Calero, 'Contesting Energy Transitions: Wind Power and Conflicts in the Isthmus of Tehuantepec' (2017) 24(1) *Journal of Political Ecology* 992; Sofia Avila, 'Environmental Justice and the Expanding Geography of Wind Power Conflicts' (2018) 13(3) *Sustainability Science* 599.

8 ProDESC, 'EDF in Mexico: Paris Court Misses Opportunity to Prevent Human Rights Violations', *ProDESC* (1 December 2021) <https://prodesc.org.mx/en/press-release-edf-in-mexico-paris-court-misses-opportunity-to-prevent-human-rights-violations>; *Case Report: Union Hidalgo v EDF* (European Centre for Constitutional and Human Rights and ProDESC, November 2020); European Centre for Constitutional and Human Rights (n 7).

experienced by people living nearby was considered not serious enough to override the benefits of shifting to renewable energy.[9]

In many countries, the transition to renewable energy has been burdened by the economic and political question of how to address the concerns of workers and communities who currently rely on fossil fuels, including the mining, energy and transportation sectors. The challenge of ensuring a just transition implicates many human rights, most notably rights to work and to fair and safe working conditions. Under the *International Covenant on Economic, Social and Cultural Rights*, countries must guarantee that everyone has an opportunity to gain a living by work of their choosing.[10] They further must guarantee fair wages, safe working conditions and reasonable working hours.[11] Limitations on work-related rights can have flow-on effects for other rights. For instance, the right to an adequate standard of living (including rights to housing and food) is clearly affected by the loss of work or a reduction in income.[12] This then potentially implicates the right to social security, which requires states to provide financial support for people in the event of unemployment.[13] There may also be consequences for freedom of movement or the right to private and family life if people are compelled to move in order to find work.[14] Managing the transition to renewable energy must therefore be done in a way which respects these human rights by avoiding large-scale negative disruption of the workforce. While these human rights concerns are first and foremost an issue for current generations, if the transition to renewable energy is not managed with care, communities, towns and regions could suffer an economic blow with intergenerational implications.

There are very serious human rights concerns associated with the extraction of critical minerals needed for renewable technology, particularly cobalt, lithium, nickel and rare earth elements. The shift to renewable energy has created an expanding market for these resources which many states are seeking to exploit, but in some cases those industries are associated with human rights violations like child exploitation, wage theft, poor occupational safety and environmental destruction.[15] Perhaps the best known example is

9 *Fagerskiold v Sweden* [2008] Eur Court HR, Application No 37664/04.
10 *International Covenant on Economic, Social and Cultural Rights* 1966 art 6 ('ICESCR').
11 Ibid art 7.
12 Ibid art 11.
13 Ibid art 9.
14 *International Covenant on Civil and Political Rights* 1966 arts 12 and 17 ('ICCPR').
15 Benjamin K Sovacool, 'Who Are the Victims of Low-Carbon Transitions? Towards a Political Ecology of Climate Change Mitigation' (2021) 73 *Energy Research & Social Science* 101916; Benjamin K Sovacool and Michael H Dworkin, 'Energy Justice: Conceptual Insights and Practical Applications' (2015) 142 *Applied Energy* 435; Lehmann and Tittor (n 7).

cobalt mining in the Democratic Republic of the Congo, where many children work in hazardous conditions where they are exposed to harmful chemicals and the risk of mines collapsing.[16] As with any extractive industry, mining of critical minerals raises human rights issues connected to environmental degradation, disruption of social and cultural connections, and Indigenous rights to land and culture.

There is also the issue of what to do with the electronic waste (e-waste) that is generated by new forms of energy production. For instance, decommissioned solar panels are already producing large volumes of e-waste which are expected to grow considerably in the future.[17] Lithium-ion rechargeable batteries – used for solar power storage and electric vehicles, bikes and scooters, as well as many common household items – have a high risk of catching fire if not stored and disposed of appropriately. Physical damage or exposure to high temperatures can cause the battery to fail, igniting the flammable substances within them. These fires are then very difficult to extinguish, because the heat of the fire triggers the battery's own chemical reactions, generating more and more heat in an escalating cycle.[18]

Currently, many countries export their e-waste for recycling, often to states in the Global South. This waste is either recycled or placed in landfill, both of which create environmental hazards. Recycling often involves burning or treating the waste products with acid to extract the raw materials, both of which generate hazardous chemicals that can be very damaging to the health of humans and the environment.[19] As the transition to renewable energy

[16] *This Is What We Die for: Human Rights Abuses in the Democratic Republic of the Congo Power the Global Trade in Cobalt* (Amnesty International, 2016) <www.amnesty.org/en/wp-content/uploads/2021/05/AFR6231832016ENGLISH.pdf>; Jonathan Watts, 'How the Race for Cobalt Risks Turning It from Miracle Metal to Deadly Chemical', *The Guardian* (online, 18 December 2019) <www.theguardian.com/global-development/2019/dec/18/how-the-race-for-cobalt-risks-turning-it-from-miracle-metal-to-deadly-chemical>; Bianca Nogrady, 'Cobalt Is Critical to the Renewable Energy Transition. How Can We Minimize Its Social and Environmental Cost?' *Ensia* (14 May 2020) <https://ensia.com/features/cobalt-sustainability-batteries>.

[17] Keshav Parajuly et al, *Future E-Waste Scenarios* (StEP [Bonn], UNU ViE-SCYCLE [Bonn] and UNEP IETC [Osaka], 2019) <https://wedocs.unep.org/bitstream/handle/20.500.11822/30809/FutEWSc.pdf?sequence=1&isAllowed=y>; CP Baldé et al, *Global Transboundary E-Waste Flows Monitor 2022* (2022).

[18] Neil Martin, 'Seven Things You Need to Know about Lithium-Ion Battery Safety', *UNSW Sites* (20 March 2023) <www.unsw.edu.au/newsroom/news/2023/03/seven-things-you-need-to-know-about-lithium-ion-battery-safety>.

[19] Benjamin K Sovacool et al, 'The Decarbonisation Divide: Contextualizing Landscapes of Low-Carbon Exploitation and Toxicity in Africa' (2020) 60 *Global Environmental Change* 102028.

accelerates, the problem of e-waste will similarly increase and, if not addressed, will add to the burden facing future generations.

7.2.3 *Geoengineering*

The profound and unprecedented environmental change that is currently unfolding is occurring concurrently with a period of rapid technological innovation. This combination of factors makes the field of environmental technology one full of exciting possibilities. Perhaps the most enticing of these is geoengineering – the deliberate manipulation of the Earth's systems – which has been proposed as a potential strategy for counteracting anthropogenic climate change.[20] The possibility of 'fixing' climate change through technology has obvious appeal. Furthermore, the scientific modelling of pathways that can succeed in keeping global warming within the target limit of 1.5°C assumes that strategies for 'negative emissions' are in place, either in the form of changes to agriculture, forestry and land use, or through carbon capture and storage.[21] There is therefore a considerable prospect that proposals for geoengineering projects will continue to come before governments and other decision-makers.

However, while geoengineering might appeal as a solution to climate change, it also raises several issues of concern for the environmental rights of future generations. These risks are present not only in the implementation of geoengineering, but also in the research and experimentation which must take place prior to its deployment. Geoengineering might have significant

[20] The ethical implications of geoengineering have been the subject of considerable scholarly debate for some time. See e.g. David R Morrow, 'Why Geoengineering Is a Public Good, Even If It Is Bad' (2014) 123(2) *Climatic Change* 95; Stephen M Gardiner, 'Some Early Ethics of Geoengineering the Climate: A Commentary on the Values of the Royal Society Report' (2011) 20(2) *Environmental Values* 163; Stephen M Gardiner, 'Why Geoengineering Is Not a "Global Public Good", and Why It Is Ethically Misleading to Frame It as One' (2013) 121(3) *Climatic Change* 513; Stephen M Gardiner, 'A Call for a Global Constitutional Convention Focused on Future Generations' (2014) 28(3) *Ethics & International Affairs* 299; Stephen Gardiner, 'Why "Global Public Good" Is a Treacherous Term, Especially for Geoengineering' (2014) 123(2) *Climatic Change* 101; Toby Svoboda, 'Is Aerosol Geoengineering Ethically Preferable to Other Climate Change Strategies?' (2012) 17(2) *Ethics and the Environment* 111; Christopher J Preston, 'Re-thinking the Unthinkable: Environmental Ethics and the Presumptive Argument against Geoengineering' (2011) 20(4) *Environmental Values* 457; Sam Adelman, 'Geoengineering: Rights, Risks and Ethics' (2017) 8(1) *Journal of Human Rights and the Environment* 119.

[21] Myles Allen et al, *Intergovernmental Panel on Climate Change Special Report on Global Warming of 1.5°C: Summary for Policymakers* (Intergovernmental Panel on Climate Change, 6 October 2018) C.1.

negative impacts on atmospheric, geologic and marine systems, with possibly severe consequences for future generations. There is also risk associated with the technology options we choose. Decisions about which technologies we research, field-test and scale up, the locations of those activities and the ways in which they will be monitored and regulated, create the potential for path dependence and lock-in which may foreclose other options for future generations.[22] And because there is still uncertainty about how (and if) geoengineering will work, the governance of geoengineering must satisfy the dual imperatives of managing potential risks while at the same time encouraging research so we can capitalise on its potential and improve our understanding of the associated dangers.[23]

To elaborate with a couple of illustrations, we can look at two examples of geoengineering technology: sulphur aerosol injection (SAI) and bioenergy with carbon capture and storage (BECCS). SAI aims to slow global warming by injecting aerosols into the atmosphere to enhance the Earth's ability to reflect the sun's heat. It is an example of a broader category of geoengineering called solar radiation management (SRM), which could counteract the greenhouse effect by reducing the amount of heat which penetrates through to the lower levels of the Earth's atmosphere.

Like all SRM technologies, SAI poses the serious risk of 'termination shock'. Because methods like SAI do nothing to address the emission of greenhouse gases, they can disguise the warming effect of those emissions while they continue to build up. If the SRM technology fails or is discontinued (e.g. due to the outbreak of war or a significant change in political attitudes or

[22] Catriona McKinnon, 'Sleepwalking into Lock-in? Avoiding Wrongs to Future People in the Governance of Solar Radiation Management Research' [2018] *Environmental Politics* 1.

[23] KA Brent and J McGee, 'The Regulation of Geoengineering: A Gathering Storm for International Climate Change Policy?' (2012) 46(4) *Air Quality and Climate Change* 22; Kerryn Brent et al, 'International Law Poses Problems for Negative Emissions Research' (2018) 8(6) *Nature Climate Change* 451; Kerryn Brent, Jeffrey McGee and Amy Maguire, 'Does the "No-Harm" Rule Have a Role in Preventing Transboundary Harm and Harm to the Global Atmospheric Commons from Geoengineering?' [2015] *Climate Law* 35; Kerryn Brent, Jeffrey McGee and Jan McDonald, 'The Governance of Geoengineering: An Emerging Challenge for International and Domestic Legal Systems?' (2015) 24(1) *Journal of Law, Information and Science* 1; Jeffrey McGee, Kerryn Brent and Wil Burns, 'Geoengineering the Oceans: An Emerging Frontier in International Climate Change Governance' (2018) 10(1) *Australian Journal of Maritime & Ocean Affairs* 67; David Morrow, 'International Governance of Climate Engineering: A Survey of Reports on Climate Engineering, 2009–2015' [2017] *SSRN Electronic Journal* <www.ssrn.com/abstract=2982392>; Carnegie Council for Ethics in International Affairs, 'Carnegie Climate Geoengineering Governance Initiative (C2G2)' <www.carnegiecouncil.org/programs/ccgg/index>.

international relations) it would lead to rapid warming of the planet with potentially catastrophic consequences.[24]

Because it involves manipulation of the Earth's atmosphere and clouds, SAI risks disrupting the hydrological cycle: the processes of cloud formation, precipitation and evaporation. Some scientific studies predict that SAI could have serious adverse effects on mean precipitation rates and rainfall patterns, and could further disrupt the African and Asian monsoons.[25] These changes would have disproportionately severe impacts on the Global South, including areas already prone to drought and extreme flooding events, and island nations, which rely on rainfall for freshwater supplies. The human rights implications of such interferences are clear. Most obviously they present threats to the human rights to food, water and the highest attainable standard of health. When these impacts are severe, they threaten the right to life. SAI may also present risks to the ozone layer, potentially undoing the progress which has been made to date under the *Montreal Protocol*.[26] Given the significant number of skin cancer deaths associated with UV exposure every year, such a setback would amount to a significant interference with the right to health[27] as well as the right to life.[28]

BECCS is an example of negative emissions technology which reduces emissions by using plants as a source of bioenergy. The plants draw carbon out of the atmosphere as they grow, and carbon capture and storage techniques are then employed in the energy generation process to limit further emissions. The human rights risks associated with BECCS flow from its massive demand for land and water. Vast areas of land are required in order to grow biofuel in the quantities needed to make a significant impact on greenhouse gas levels. The reallocation of land from agriculture threatens food security and disruption of livelihoods, and incentives for governments to take up BECCS may lead to forced displacement of people from their land.[29] Proponents argue that land could be sourced which is currently under-utilised,[30] but this creates

[24] McKinnon (n 22); William CG Burns, *The Paris Agreement and Climate Geoengineering Governance: The Need for a Human Rights-Based Component* (No 111, Centre for International Governance Innovation, October 2016) 14; Brent and McGee (n 23) 25; Seth D Baum, 'The Great Downside Dilemma for Risky Emerging Technologies' (2014) 89(12) *Physica Scripta* 128004; Andrew Ross and H Damon Matthews, 'Climate Engineering and the Risk of Rapid Climate Change' (2009) 4(4) *Environmental Research Letters* 045103.

[25] Burns (n 24); Royal Society (Great Britain), *Geoengineering the Climate: Science, Governance and Uncertainty* (The Royal Society, 2009).

[26] Burns (n 24) 13; Brent and McGee (n 23) 25.

[27] *ICESCR* (n 10) art 12.

[28] *ICCPR* (n 14) art 6.

[29] Burns (n 24) 16.

[30] Ibid 16–17.

other challenges, particularly where such areas are the traditional lands of Indigenous peoples.

In addition to the ways in which these technologies might limit the ability of future generations to enjoy their human rights, there are also concerns relating to the way that decisions around geoengineering are made. Adelman has argued strongly that the potential for widespread and significant negative impacts demands that any decision to proceed with geoengineering be made with the consent of those likely to be affected.[31] However, he has also articulated several significant problems with consent in this context. One of the key challenges is guaranteeing that consent would be freely obtained and adequately informed, especially given the imbalance of power which exists between those conducting the research or deployment (mostly European and North American scientists, mostly male) and those likely to be affected (mostly vulnerable communities from the Global South).[32] When the affected groups are future generations, we can add the question of how consent is to be meaningfully obtained on behalf of groups who lack a voice in the debate.

7.2.4 *Adaptation*

Just as there are human rights risks associated with our actions to reduce net greenhouse gas emissions, there are also potential issues arising from climate adaptation policies, that is, the strategies we put in place to adapt to eventual climate change and minimise its harmful effects. Climate adaptation comes in a wide range of forms, as particular strategies need to be tailored to the communities and sectors whose needs they address. For example, in the agricultural sector, adaptation might involve changing practices or growing different crops that are better suited to a changed climate. Water management practices are also likely to need to adapt to changes in rainfall patterns. In many places, public transport schemes will need to adapt to higher temperatures which will impact road and rail networks and the people who need to access them.

The increasing frequency and severity of disasters like floods, fires, cyclones and heatwaves demands considerable planning and investment in adaptation. New building codes might be needed to ensure housing is designed and constructed to better withstand disasters and manage extreme heat. Planning regimes might need to be changed to avoid construction in areas that are at higher risk of flooding or fire. Laws which regulate tenancies and community

[31] Adelman (n 20).

[32] Ibid 134; Mike Hulme, *Can Science Fix Climate Change? A Case against Climate Engineering* (Polity Press, 2014) 54.

title schemes will need to adapt to ensure housing is suitable and responsibilities are fairly shared between owners and tenants.

For low-lying states and coastal communities, adaptation will involve all of the above, but potentially also the construction of protective infrastructure like sea walls. In the worst cases, adaptation may need to take the form of partial or total relocation of communities away from low-lying areas, or even the full movement of a population to another island or migration to another state.

Each of these strategies has the potential to impact on human rights, especially where they are implemented without adequate consultation and participation of the people affected. Many of them will also have consequences for future generations, both beneficial and potentially harmful. The future consequences of adaptation planning must therefore be carefully considered. However, the risk for future generations arises mostly where states have failed to plan for or implement adequate adaptation strategies. Delays in taking adaptation action create an increased burden for future generations. As the next section describes, the failure to implement adaptation and protect human rights from climate change has been a theme in a number of climate litigation cases.

7.3 CLIMATE LITIGATION TO DATE

This section provides a brief description of some of the key cases which have sought to address intergenerational impacts of climate change and highlights some common challenges as well as the features of cases that have been successful. It complements the discussion of case law presented in Chapter 3, which examined the latest developments in jurisprudence relating to the environmental rights of future generations and intergenerational equity. This section concentrates on climate change specifically, however, with a particular focus on litigation by children and young people.

As at the end of December 2022, thirty-four human rights-based cases had been commenced by children or young people in domestic, regional or international courts and tribunals.[33] A number of new cases were also filed in 2023, with some significant decisions handed down in the past few years.

[33] Michael Burger and Maria Antonia Tigre, *Global Climate Litigation Report: 2023 Status Review* (Sabin Centre for Climate Change Law, Columbia Law School and United Nations Environment Programme, July 2023) 40; see also Joana Setzer and Catherine Higham, *Global Trends in Climate Change Litigation: 2024 Snapshot* (Grantham Research Institute on Climate Change and the Environment, London School of Economics and Political Science, 2024) <www.lse.ac.uk/granthaminstitute/wp-content/uploads/2024/06/Global-trends-in-climate-change-litigation_2024-snapshot.pdf>; Bridget Lewis, 'Children's Human Rights-Based Climate Litigation at the Frontiers of Environmental and Children's Rights' (2021) 39(2) *Nordic Journal of Human Rights* 180.

7.3.1 *Future Generations v Ministry for the Environment of Colombia (2018)*

The decision of the Colombian Supreme Court in *Demanda Generaciones Futuras v Minambiente* was a significant milestone in climate litigation for future generations.[34] Unlike some of the other cases discussed here, which include intergenerational considerations in addressing the rights of children and young people, this case was explicitly brought on behalf of future generations.

Twenty-five young people aged seven to twenty-five years brought a claim against the government for failing to address climate change, particularly by allowing deforestation of the Amazon rainforest. The court held that the government's failure to protect the Amazon rainforest and the resulting contribution to global heating constituted a breach of the right to a healthy environment, which was interpreted broadly to extend to non-human species and ecosystems.[35]

7.3.2 *Pandey v India (2019)*

In India, Ridhima Pandey, then aged nine years old, brought a complaint to the National Green Tribunal, which hears cases relating to matters of significant environmental concern. She argued that India was obliged to adopt stronger climate mitigation policies based on the public trust doctrine and India's commitment to the Paris Agreement. She also argued that India's environmental laws include a duty to consider climate impacts, since any consideration of the environment must implicitly include the climate as well.

The petition describes children and future generations as a class of people who possess a right to a healthy environment, but who are especially vulnerable to climate change. It relied on principles of intergenerational equity and sustainable development, pointing to the fact that climate change would impact disproportionately on children and future generations. It further argued that children and future generations have no meaningful way to protect themselves against the consequences of India's policies.

The National Green Tribunal dismissed the claim in a two-page decision. Noting the obligation in India's environment legislation for environmental impacts to be assessed, the tribunal ruled that this would necessarily include

[34] *Demanda Generaciones Futuras v Minambiente* [2018] (11001-22-03-000-2018-00319-01, 5 April 2018).

[35] Ibid [34].

an assessment of climate impacts. It said: 'There is no reason to presume that Paris Agreement and other international protocols are not reflected in the policies of the Government of India or are not taken into consideration in granting environment clearances'.[36] On this basis, the tribunal was of the view that there was no question which needed to be determined, and the case was dismissed.

7.3.3 *Neubauer v Germany (2021)*

In 2021, the German Federal Constitutional Court handed down its decision in *Neubauer v Germany*.[37] Luisa Neubauer is a young climate activist, and one of the architects of the 'Fridays for Future' school climate strike movement.[38] She brought a legal action against Germany arguing that the government's failure to implement stronger emissions reduction policies breached the state's obligation to safeguard the rights to life and the environment, protected under the German constitution. The court upheld her complaint, ruling that the legislature must establish targets and measures that are capable of meeting the objectives of the Paris Agreement. It also stressed that the responsibility for reducing emissions must be shared fairly between current and future generations – a failure to implement strong policies now would present an unfair burden for future generations.

7.3.4 *Sacchi et al v Argentina et al (2021)*

Another of the Fridays for Future leaders, well-known climate campaigner Greta Thunberg, was one of the claimants in the case of *Sacchi et al v Argentina et al*.[39] This was a complaint to the UN Committee on the Rights of the Child, brought by sixteen children against five states – Argentina, Brazil, France, Germany and Turkey. These five states were selected because they

[36] *Pandey v Union of India* [2019] National Green Tribunal, Principal Bench, New Delhi 187/2017, 2.

[37] *Neubauer et al v Germany* [2021] Federal Constitutional Court 1 BvR 2656/18, 1 BvR 78/20, 1 BvR 96/20, 1 BvR 288/20.

[38] 'FridaysForFuture' <www.fridaysforfuture.org>.

[39] *Sacchi et al v Argentina, Brazil, France, Germany and Turkey* (Communication to the Committee on the Rights of the Child, Sacchi Chiara et al, 23 December 2019) ('*Sacchi et al (Petition)*'); *Sacchi et al v Argentina* [2021] Committee on the Rights of the Child Communication No 104/2019; for further analysis see Bridget Lewis, 'Sacchi et al v Argentina, Brazil, France, Germany and Turkey (United Nations Committee on the Rights of the Child)' (2022) 36(9/10) *Australian Environment Review* 201.

were parties to both the Convention on the Rights of the Child and the Paris Agreement, but were not on track to meet their emissions reduction targets.

The children claimed that the respondents' climate policies violated their rights under the CRC, in particular the rights to life (article 6), health (article 24) and culture (article 30). Their complaint covered a range of allegations, reflecting the diverse ways that climate change has already impacted them and their communities. For example, children from Indigenous communities in Sweden and Alaska pointed to the impact of global warming on traditional practices of hunting, fishing and reindeer herding. Children from the Marshall Islands argued that ocean warming was having a detrimental effect on fishing, at the same time that rising sea levels and storm surges threaten lives and livelihoods.

One of the key issues in the case was demonstrating that the five respondent states owed legal obligations to protect the children's rights, given that most of the children came from other countries. As explained in Chapter 5, the committee did not see this as a barrier, however. Instead, it declared that states owe human rights duties beyond their borders where activities that they control have foreseeable extraterritorial human rights impacts. Applying that test, the committee held that the respondent states owed duties to the children under the CRC because the global consequences of their inadequate climate policies were foreseeable.

That being said, the committee decided that the complaint was inadmissible because the children had not tried to resolve their dispute through domestic channels first, which is a threshold requirement under the Optional Protocol which governs the complaints process.[40] While the substantive question of whether inadequate emissions reduction represented a breach of human rights was therefore not considered, the case still represented a major milestone in international human rights-based climate litigation.

7.3.5 *Daniel Billy v Australia (2022)*

The following year, another UN-based climate case was concluded, this time in the Human Rights Committee, which hears complaints under the International Covenant on Civil and Political Rights. In *Daniel Billy v Australia*, a group of Torres Strait Islanders argued that the Australian government had failed to do enough to protect their islands and their way of life from

[40] *Optional Protocol to the Convention on the Rights of the Child on a Communications Procedure* 2011 art 7(e).

the impacts of climate change.[41] The complainants were eight adults, but they also brought the case on behalf of six of their children.

They alleged that Australia had violated their rights to life, culture and private and family life through its inadequate emissions reduction policies and its failure to adequately protect against rising sea levels. Torres Strait Islanders are experiencing saltwater inundation of their lands, leading to destruction of homes, gardens, ancestral burial lands and other culturally significant sites. Freshwater supplies are being contaminated by seawater, and culturally and nutritionally important marine species have been detrimentally affected by ocean warming. The cultural ties between Torres Strait Islanders and their lands mean that relocation to other areas is not an option.

The committee held that Australia had breached its obligation to take reasonable measures to protect Torres Strait Islanders against foreseeable threats to their human rights. It found violations of the right to family life and the right to culture, though the majority did not consider that the threat was yet imminent enough to constitute a violation of the right to life. Similarly, the committee was reluctant to engage with the intergenerational dimensions of the petitioners' claim, instead focusing on Australia's need to take action to protect against present threats.

7.3.6 *Waratah Coal v Youth Verdict (2022)*

Another Australian case decided in 2022 represented a significant step forward in terms of protecting the rights of future generations from climate change. The decision in *Waratah Coal v Youth Verdict* gave greater weight to intergenerational issues in examining the human rights implications of a potential new coal mine in the Galilee Basin in central Queensland.[42] The objectors to the mine included a group of young people (the organisation Youth Verdict) as well as the custodians of a nature refuge which overlapped with the footprint of the mine (the Bimblebox Alliance). They argued that approval of the mine would be incompatible with human rights protected under Queensland's *Human Rights Act 2019*.

In deciding whether to recommend approval of the mine, President Kingham of the Land Court considered the contribution the proposed mine would make to climate change after the coal was exported and burned

[41] *Daniel Billy and Others v Australia* [2022] Human Rights Committee Communication No. 3624/2019.

[42] *Waratah Coal Pty Ltd v Youth Verdict Ltd and Others (No 6)* [2022] Land Court of Queensland QLC, 21, 1339.

overseas. She concluded that the future environmental damage would negatively impact on Queenslanders' rights to life, property, private and family life, and the cultural rights of Aboriginal and Torres Strait Islander peoples, which are explicitly protected under Queensland's human rights law. She also judged that these impacts could not be justified by the economic benefits the mine would bring.

Thinking about the impacts on children and future generations, President Kingham noted that the global warming caused by the mine would disproportionately impact on children and future generations. She also gave weight to the intergenerational dimension of Aboriginal and Torres Strait Islander peoples' cultural rights, which include the right to maintain culture and language through passing these on to younger generations.

These intergenerational considerations were explained in the testimony of many First Nations witnesses in the case. As one witness, Juritju Fourmile, explained:

> We as Indigenous people, First Nations people don't just think about one generation or ourselves. We think about the generations that come after us, those that come before us as well . . . Climate change is forcing us off of our traditional homelands now and the events that are happening now. Where is my daughter going to go fishing later on when all the creeks dry up? Where is she going to go when she can't go into the bush because it's too hot and the rainforest dies off? She can't then pass those stories down to the next generation.

The case gave legal recognition to the fact that, especially with respect to Aboriginal and Torres Strait Islander culture, the future impacts of climate change represent a present violation of rights, since those rights include the ability to transmit culture and the sense of responsibility to future generations. It also recognised the need to consider the full extent of the emissions that flow from a fossil fuel project such as that proposed by Waratah Coal. In much of the Australian jurisprudence, courts had been reluctant to place weight on the scope 3 emissions – the greenhouse gases that would be emitted from the eventual burning of the coal once sold. President Kingham rejected the notion that these 'downstream' emissions are not to be taken into account, noting that '[t]his case is about Queensland coal, mined in Queensland, and exported from Queensland to be burnt in power stations to generate electricity. Wherever the coal is burnt the emissions will contribute to environmental harm, including in Queensland'.[43] The foreseeable human rights

[43] Ibid [26].

consequences of scope 3 emissions must therefore be considered and, in this case, could not be justified.

7.3.7 *Sharma v Minister for Environment* (2022)

Australia's courts have not always been welcoming of claims made on the basis of intergenerational equity. In *Sharma v Minister for Environment*, a high school student, Anjali Sharma, brought an action against the federal environment minister's decision to approve an extension of a coal mine. Using the law of negligence, she argued that the Minister for Environment has a duty of care to protect children from climate harms when considering whether to approve fossil fuel extraction.

At first instance, Justice Bromberg of the Federal Court agreed that the Minister had such a duty of care.[44] He found that scope 3 emissions from the proposed coal mine expansion would make a 'small but measurable increase' to global average surface temperatures and present a real risk that the young plaintiffs would face harm from associated climatic hazards. Applying the common law of negligence, he concluded that a reasonable person in the Minister's position would foresee that the children are 'exposed to a risk of death or personal injury' as a result of the operation of the mine project. Consequently, the Minister owes a duty to take reasonable care to avoid injury to the children. The children's vulnerability to climate change and the fact that they rely on the Minister to prevent climatic harms were key factors in the court's finding that a duty of care exists.

However, this decision was overturned on appeal in the Full Court of the Federal Court of Australia.[45] The appeal was heard by three judges, each of whom gave separate reasons for overturning the original decision. However, the appeal judges were unanimous in that they did not overrule any of the primary judge's findings with respect to the nature of the threat which climate change poses to children. Overall, the appeal judges concluded that there wasn't a sufficiently close relationship between the Minister's decision and the serious risks of climate change to give rise to a common law duty of care. The judges also felt that the question of how the Minister should regard the future risks of climate change is a matter of policy-making which should not be determined by the judiciary.

Since the decision, Anjali Sharma has worked closely with an independent Australian senator, David Pocock, to introduce legislation which would

[44] *Sharma by her litigation representative Sister Marie Brigid Arthur v Minister for the Environment* (2021) [2021] FCA 560 (Federal Court of Australia, 27 May 2021).

[45] *Minister for the Environment v Sharma* (2022) [2022] FCAFC 35 (15 March 2022).

address the gap left by the Sharma appeal. The *Climate Change Amendment (Duty of Care and Intergenerational Climate Equity) Bill* was introduced in 2023. The Bill proposed the addition of a statutory duty to the *Climate Change Act* requiring decision-makers to consider the health and well-being of current and future children when determining whether to greenlight a fossil fuel project.[46] It also sought to introduce a prohibition on any decision that would result in large volumes of greenhouse gases that would pose a 'material risk of harm' to the health and well-being of children, either now or in the future.[47]

The Bill was considered at length by a parliamentary committee who received over 400 submissions, the majority of which supported stronger legal provisions for intergenerational climate justice. Ultimately the committee decided not to recommend the Bill, but it did acknowledge that climate change represents a significant threat to the well-being of children and future generations.[48] Despite this setback, advocates for intergenerational climate justice continue to work for stronger legal protections.[49]

7.3.8 *ENvironnement JEUnesse v Canada* (2022)

Around the same time in Canada, another case brought by young people was making its way through the courts. The organisation ENvironnement JEUnesse sought to bring a class action on behalf of people under the age of thirty-five in Québec, arguing that Canada's policies on climate change constituted a breach of their human rights under both the Canadian and the Québec Charters of Rights and Freedoms. Specifically, they argued that inadequate emissions reduction targets represented a breach of the right to life under both human rights instruments and the specific right to a healthy environment recognised in the Québec charter.[50]

The complaint was initially dismissed by the Superior Court of Québec. While the judge determined that the impacts of climate change on human rights was a justiciable matter, he concluded that the cut-off of thirty-five years

[46] *Climate Change Amendment (Duty of Care and Intergenerational Climate Equity) Bill 2023* s 15D.

[47] Ibid s 15E.

[48] Senate Standing Committees on Environment and Communications, Report of Inquiry into Climate Change Amendment (Duty of Care and Intergenerational Climate Equity) Bill 2023 (Parliament of Australia, June 2024).

[49] 'Parliamentary Friends for Future Generations – EveryGen' (27 March 2023) <www.everygen.online/2023/03/27/parliamentary-friends-for-future-generations>.

[50] *ENvironnement JEUnesse v Procureur General du Canada (Application)* (Superior Court of Québec, 26 November 2018).

of age was arbitrary and could not be the basis for a class action.[51] The applicants appealed to the Québec Court of Appeal but were again dismissed. On appeal, the court ruled that the adequacy of Canada's climate policies and the question of what those policies should entail is a matter for the legislature, not the judiciary, to determine.[52] ENvironnement JEUnesse sought leave to appeal to the Canadian Supreme Court, but this was denied in 2022.

7.3.9 *Held v State of Montana (2023)*

One of the most recent cases to explore the environmental rights impacts of climate change for young people and future generations was *Held v Montana*.[53] A group of young plaintiffs argued that Montana's fossil fuel-based energy policy violated their constitutional rights. In particular, they challenged a requirement of the *Montana Environmental Policy Act* (MEPA) which prohibited consideration of emissions and resulting climate change in environmental decision-making.

Kathy Seeley of the First Judicial District Court accepted that Montana's climate is warming at an increasing rate and that this is a direct consequence of the state's contribution to global greenhouse gases. She also accepted that children are particularly vulnerable to the effects of climate change. Judge Seeley found that the provision in the MEPA has permitted fossil fuel projects to go ahead despite the state having knowledge that these would cause detrimental impacts to young people and other residents of Montana.

The plaintiffs argued that this represented a violation of their constitutional rights, including their right to a clean and healthful environment, which is listed among the inalienable rights guaranteed to all persons.[54] This right is also supported by a duty placed on the state and all people to 'maintain and improve a clean and healthful environment in Montana for present and future generations'.[55] The constitution also specifies that people under the age of eighteen are entitled to all the same rights as adults.[56] Judge Seeley agreed, ruling that the provision in MEPA was unconstitutional and ordering that the

51 *ENvironnement JEUnesse v Procureur General du Canada* [2019] Superior Court of Québec 500-06-000955-183.

52 *ENvironnement JEUnesse v Canada* [2021] QCCA Quebec Court of Appeal 500-09-028523-199, 1871.

53 *Held et al v State of Montana (Order)* [2023] Montana First Judicial District Court (Lewis and Clark County) Cause No CDV-2020-307; *Held et al v State of Montana (Opinion)* [2024] Supreme Court of Montana (Case No DA 23-0575) 2024 MT 312.

54 *The Constitution of the State of Montana (Annotated)* 2023 art II, s 3.

55 Ibid art IX s 1.

56 Ibid art II s 15.

state was prohibited from applying it. The State of Montana appealed the District Court's decision, arguing that the constitutional provisions which the children relied on are not self-executing.[57] In a 6-1 majority opinion, the Supreme Court confirmed that the young plaintiffs had standing and upheld the District Court's ruling that the MEPA provision breached Montanans' right to a clean and healthy environment.

7.3.10 *Duarte Agostinho and Others v Portugal and Thirty-Two Others* (2024)

Duarte Agostinho et al v Portugal et al was one of several highly anticipated cases to be decided by the European Court of Human Rights. Six young people from Portugal brought a case against thirty-three European states, alleging that their collective failure to reduce emissions had resulted in harm to their human rights.[58] They claimed that climate change violates their rights to life and private and family life under the European Convention. In particular, they pointed to the heat waves and bushfires which Portugal has experienced over recent years, arguing that these demonstrate that climate change is not only a reasonably likely future occurrence, but is already having serious impacts on the health and well-being of young people.

As with *Sacchi v Argentina*, there were issues of admissibility relating to both extraterritorial jurisdiction and exhaustion of local remedies. With respect to extraterritorial obligations, the claimants adopted a similar approach to that taken in *Sacchi*, arguing that obligations were owed to them based on the control exercised by the states over their emissions policies together with the foreseeability of the resulting global warming.[59] With respect to the issue of exhaustion of local remedies, they argued that the urgency of climate change meant that there was no time to pursue multiple domestic proceedings.

After the complaint was lodged, the European Court moved the case to its Grand Chamber in order to fast-track the hearing, which was taken by some to be a sign that the court may have been positively disposed to these arguments.[60] However, in its judgment of April 2024, the court found that there were no

[57] *Held et al v State of Montana* (Notice of Appeal) [2023] Supreme Court of Montana DA 23-0575.

[58] *Cláudia Duarte Agostinho and Others v Portugal and 32 Other States* (Eur Court HR, Application No 29371/20).

[59] Ibid [18]; citing *Andreou v Turkey* [2010] Eur Court HR, Application No 45653/99; and *Kovačić et al v Slovenia* [2008] Eur Court HR 44574/98, 45133/98 and 48316/99.

[60] For more discussion, see Ole W Pedersen, 'The European Convention of Human Rights and Climate Change – Finally!', *EJIL: Talk!* (22 September 2020) <www.ejiltalk.org/the-european-convention-of-human-rights-and-climate-change-finally>; Lewis, 'Children's

grounds to establish extraterritorial obligations and that the case was inadmissible owing to the failure to exhaust local remedies. As with *Sacchi*, the case did not proceed to address the substantive questions of how the rights in the European Convention must be protected in the context of climate change.

7.3.11 *Verein KlimaSeniorinnen Schweiz v Switzerland* (2024)

Verein KlimaSeniorinnen Schweiz v Switzerland was one of two other decisions handed down by the European Court of Human Rights at the same time as *Duarte Agostinho*. The claimants were the not-for-profit organisation KlimaSeniorinnen Schweiz (Senior Women for Climate Protection Switzerland) along with four of its individual members. While the case is distinct from those discussed earlier owing to the age of the claimants, it has perhaps been the most significant human rights decision in recent decades in terms of advancing intergenerational equity.

Chief among its historic findings was the way in which the court redefined the tests for victim status and standing in climate cases. As discussed in more detail in Chapter 4, the court found that an organisation could have standing to represent the interests of affected individuals who faced threats or adverse effects of climate change on their life, health and well-being.[61] In relation to future generations, the court recognised that climate change raises particular issues of intergenerational burden-sharing.

> While the legal obligations arising for states under the Convention extend to those individuals currently alive who, at a given time, fall within the jurisdiction of a given Contracting Party, it is clear that future generations are likely to bear an increasingly severe burden of the consequences of present failures and omissions to combat climate change and that, at the same time, they have no possibility of participating in the relevant current decision-making processes.[62]

In considering the merits of the case, the court differentiated between states' obligations under article 2 (the right to life) and article 8 (the right to respect for private, family and home life).

As noted in Chapter 3, the European Court has a long history of finding breaches of article 8 in relation to environmental harm. Against this backdrop,

Human Rights-Based Climate Litigation at the Frontiers of Environmental and Children's Rights' (n 33).

[61] *Verein KlimaSeniorinnen Schweiz v Switzerland* [2024] Eur Court HR, Application No 53600/20 [524].

[62] Ibid [420].

the court recognised that climate change has potential to cause very profound impacts on people's well-being and their quality of life. It defined the state's duty under article 8 as requiring it 'to adopt, and to effectively apply in practice, regulations and measures capable of mitigating the existing and potentially irreversible, future effects of climate change'.[63]

The court had regard to the obligations in the UNFCCC and relevant climate science in finding that this obligation effectively requires a regulatory framework which is capable of reaching net zero emissions within three decades.[64] 'Moreover, in order for this to be genuinely feasible, and to avoid a disproportionate burden on future generations, immediate action needs to be taken and adequate intermediate reduction goals must be set for the period leading to net neutrality.'[65] Having thus defined Switzerland's obligations under the European Convention, the court held that its current emissions reduction policies failed to meet the required standards, and therefore constituted a violation of article 8.[66]

The arguments under article 2 were less successful. The court explained that states have obligations to protect against threats to life under article 2 but, in order for these obligations to be triggered, there must be a 'real and imminent' risk to life.[67] As discussed in Chapter 4, a 'real' risk is one which is serious, genuine and sufficiently ascertainable, while 'imminence' involves considerations of both physical and temporal proximity.[68] While the court recognised that the risks posed by climate change could be life-threatening and could certainly diminish quality of life, it said that whether this threat rose to the level of a 'real and imminent' risk to life was 'questionable'.[69] Having already found that a breach of article 8 had occurred, the court did not feel it was necessary to consider article 2 any further.

7.3.12 *Korean Climate Crisis Cases (2024)*

In August 2024, the Constitutional Court of South Korea handed down its decision in the 'Climate Crisis Cases'. This decision was the culmination of

63 Ibid [545].
64 Ibid [548].
65 Ibid [549].
66 Ibid [574].
67 Ibid [511].
68 Ibid [512]; *Fadeyeva v Russia* [2005] Eur Court HR, Application No 55723/00; *Brincat and Others v Malta* [2014] Eur Court HR 60908/11, 62110/11, 62129/11, 62312/11, 62338/11; *Kolyadenko and Others v Russia* [2012] Eur Ct HR, Application Nos 17423/05, 20534/05, 20678/05, 23263/05, 24283/05 and 35673/05.
69 *Verein KlimaSeniorinnen Schweiz v Switzerland* (n 61) [536].

four separate cases that had been combined into one matter. In total, they involved more than 200 plaintiffs, including about sixty children.[70] They argued that Korea's NDCs and the *Carbon Neutrality Act*, which was designed to implement the NDCs, were inadequate to protect citizens' constitutional rights. In particular, they argued violations of the rights to life, to pursue happiness, to property, to a healthy environment and to enjoy general freedom.[71] They also alleged breaches of the state's obligation to protect citizens from disasters.[72]

While South Korea had set a long-range target for net zero emissions by 2050 and a short-term target for emissions reductions by 2030, no legislation was in place to cover the period between 2031 and 2049. The Constitutional Court concluded that the failure to set interim targets shifted an excessive burden to the future to achieve emissions reductions, thereby violating the constitutional principle that prohibits insufficient protection of rights.

In the context of these four cases, the National Human Rights Commission of Korea, the state's independent human rights body, presented an opinion to the Constitutional Court which concluded that the relevant laws fail to provide minimum measures to protect the petitioners' rights.[73] The opinion notes that stronger actions are needed to protect the rights of both current and future generations. In the Commission's view, 'the current NDCs transfer a severe burden to future generations, violating intergenerational equity'.[74]

7.3.13 *Other Litigation*

In addition to the cases discussed earlier, several other cases are still before the courts at the time of writing. One of the longest-running and most publicised cases is *Juliana v US*. In 2015, a group of young people brought a constitutional case against the US government.[75] They alleged that the

70 *Woodpecker et al v South Korea* (Constitutional Court of South Korea, 2022Hunma864); *Do-Hyun Kim et al v South Korea* (Constitutional Court of South Korea); *Byung-In Kim et al v South Korea* (Constitutional Court of South Korea, 2021Hunma1264); *Min-A Park v South Korea* (Constitutional Court of South Korea, 2023Hunma846).

71 *Constitution of the Republic of Korea* 1987 arts 23, 34, 35 and 37.

72 Ibid art 34(6).

73 *Submission of Opinion to the Constitutional Court of Korea Regarding Article 8, Section 1 of the Framework Act on Carbon Neutrality and Green Growth for Coping with Climate Crisis and Article 3, Section 1 of the Enforcement Decree of the Act* (National Human Rights Commission of Korea, 12 June 2023) 1.

74 Ibid 3.

75 *Juliana v US (Complaint)* (US District Court (District of Oregon, Eugene Division), Case No 6:15-cv-01517-TC).

government's actions had contributed to climate change, and that this violated their constitutional right to life, liberty and property. They also argued that the United States has breached the public trust through its contributions to climate change. The US government has aggressively attempted to have the case dismissed, including by filing seven applications for a writ of mandamus which would compel the District Court to dismiss the case. The US argues that the young people lack standing to bring the case, and that their petition does not make out a breach of either the Constitution or the public trust doctrine. The most recent of these was filed in February 2024, and at the time of writing the Court of Appeal is considering whether to grant the motion of mandamus.[76]

In 2016, Rabab Ali, then seven years old, brought a petition against the Pakistan government seeking to stop the development of the Thar coal fields.[77] Rabab alleged that exploitation of the coal resources would violate constitutional rights and the public trust doctrine in relation to Pakistan's climate and atmosphere. No outcome has yet been handed down in the case.

In Australia, the Federal Court is currently considering its decision in *Pabai Pabai and Guy Paul Kabai v Commonwealth*. The applicants are Elders of the Gudamalulgal Nation in the Torres Strait Islands. They allege that the Australian government has a duty of care to protect their people and lands from the impacts of climate change, which it has breached by not cutting emissions rapidly enough. They point to the impacts of rising sea levels on their way of life (known as 'Ailan Kastom'), their homes, gardens, marine resources and burial sites. The government argues, in line with the finding on appeal in *Sharma v Minister for Environment* (discussed earlier), that the question of whether it owes a duty of care in relation to these impacts of climate change is one of public policy, and not to be determined by the courts.[78] At the time of writing the court's judgment is pending.

Several cases are proceeding in the European Court of Human Rights. In the case of *Greenpeace Nordic and Others v Norway*, multiple non-governmental organisations (NGOs) and six young climate activists lodged a complaint against Norway.[79] They argue that Norway's ongoing oil

[76] *US v Juliana (Petition)* [2024] Court of Appeals 9th Cir No 24-684; *US v Juliana* [2024] US Court of Appeals 9th Cir No 24-684.

[77] *Ali v Pakistan* [2016] Supreme Court of Pakistan Constitutional Petition No ___ / I of 2016.

[78] *Pabai Pabai & Anor v Commonwealth of Australia* (Federal Court of Australia, VID622/2021); Federal Court of Australia, 'Pabai Pabai & Anor v Commonwealth of Australia: Online File' (18 January 2022) <www.fedcourt.gov.au/services/access-to-files-and-transcripts/online-files/pabai-v-australia>.

[79] *Greenpeace Nordic and Others v Norway (Petition)* (Eur Court HR, 15 June 2021).

exploitation infringes upon their fundamental rights under the European Convention on Human Rights.

In *Soubeste and Others v Austria and 11 Other States*, a group of young European citizens argue that their fundamental human rights have been negatively impacted by climate change.[80] Specifically, they allege that climate change is impinging on their rights to life, to be free of inhuman or degrading treatment, to respect for their private and family life and not to be subjected to discrimination. They point to the activities of the fossil fuel industry as the source of these interferences, and allege that the respondent states have promoted this through the Energy Charter Treaty, which protects investments in fossil fuels and prevents the transition to renewable energy.

At the international level, three advisory opinions have been requested from different courts which have potential to clarify the way in which international law promotes and protects the environmental rights of future generations. The first opinion to be handed down was that of the International Tribunal for the Law of the Sea (ITLOS).[81] A group of small island states known as the Commission of Small Island States on Climate Change and International Law (COSIS) had requested ITLOS to clarify the nature of states' obligations under the *United Nations Convention on the Law of the Sea* (UNCLOS) with respect to climate change. In particular, they sought advice on the legal duties relating to ocean warming and acidification, especially for fragile marine systems like coral reefs.

The tribunal began by explaining that greenhouse gas emissions fall within the definition of 'pollution' in UNCLOS, as a substance with 'deleterious effects'. Having established this, the tribunal confirmed that states have an obligation to prevent, reduce and control marine pollution, and must therefore take the necessary means to reduce greenhouse gas emissions. ITLOS advised that what constitutes 'necessary measures' will be determined based on various factors relevant to each country's circumstances, but is not entirely at the whim of governments. The tribunal also articulated an obligation of 'stringent due diligence' that will apply in decision-making in relation to greenhouse gas emissions. While the advisory opinion does not specifically address human rights or future generations, it is an important moment in international climate law. The opinion is founded on an acceptance of the significant future harm that is likely if global emissions are not reduced, and

[80] *Soubeste and Others v Austria and 11 Other States* (Eur Court HR, No 31925/22).

[81] *Advisory Opinion on Request submitted to the Tribunal by the Commission of Small Island States on Climate Change and International Law* [2024] International Tribunal for the Law of the Sea Case No 31.

which seeks to clarify states' duties with regard to risk assessment and preventative actions. It therefore has potential to influence the law in a way which is beneficial for future generations.

The other two advisory opinions are pending at the time of writing but both have potentially more likelihood of addressing future generations explicitly. In the inter-American human rights system, Colombia and Chile have requested an advisory opinion from the Inter-American Court of Human Rights.[82] The two countries are seeking clarification from the court on states' duties of prevention in relation to the human rights impacts of climate change. In particular, the request asks about the differentiated obligations that states have towards children and the 'new generations' in the light of climate change.

In the International Court of Justice, an advisory opinion has been requested by the General Assembly following a petition by Vanuatu, which was co-signed by many other states.[83] The request asks, among other things: 'What are the obligations of States under international law to ensure the protection of the climate system and other parts of the environment from anthropogenic emissions of greenhouse gases for states and for present and future generations?' It also asks about the legal consequences that would arise if states have caused significant harm to the climate system or environment with respect to other states and to individuals of present and future generations.

While these proceedings are only advisory in nature, they do have potential to offer strong statements about the nature of states' duties with respect to the long-term impacts of climate change. In the case of the ICJ and inter-American proceedings, it is anticipated that the courts will provide welcome clarity on the scope of states' duties with respect to future generations and, as an implied consequence, on the nature of future generations' environmental rights.

7.4 LESSONS FROM THE CASES

An analysis of the cases mentioned earlier shows that, while children's climate litigation can be successful, there are certain factors that make success much more likely.

82 *Request for an advisory opinion on the Climate Emergency and Human Rights submitted to the Inter-American Court of Human Rights by the Republic of Colombia and the Republic of Chile* (Inter-American Court of Human Rights, 9 January 2023); Maria Antonia Tigre et al, 'A Request for an Advisory Opinion at the Inter-American Court of Human Rights: Initial Reactions', *Climate Law Blog* (17 February 2023) <https://blogs.law.columbia.edu/climatechange/2023/02/17/a-request-for-an-advisory-opinion-at-the-inter-american-court-of-human-rights-initial-reactions>.

83 *Obligations of States in Respect of Climate Change (Request for Advisory Opinion)* (International Court of Justice, 29 March 2023).

To begin with, successful cases have almost all been within domestic legal systems. Those that have succeeded in international or regional bodies have been complaints by citizens against their own governments (e.g. *Daniel Billy v Australia* in the Human Rights Committee and *KlimaSeniorinnen v Switzerland* in the European Court of Human Rights).

Cases have a much greater chance of succeeding where the harm is either already occurring or is imminent. In *Billy*, the Human Rights Committee emphasised Australia's duties regarding adaptation measures in the Torres Strait to protect against storm surges and rising sea levels that are already impacting the communities there. By contrast, in cases where the risk to rights is less certain or less imminent, courts have been less willing to find a breach of rights. This has been most noticeable in relation to the right to life, for example, in cases like *Teitiota v NZ* (discussed in Chapter 4) and *KlimaSeniorinnen*. However, in that latter decision the European Court emphasised the need to act quickly to avoid severe, irreversible impacts to other human rights, perhaps indicating a growing willingness to recognise positive duties to protect against foreseeable effects of climate change.

Another feature that increases the likelihood of success appears to be the existence of a specific constitutional provision upon which the case can be based, either one that creates duties with respect to future generations or one which protects environmental rights, or a combination of both. For example, both *Future Generations v Colombia* and *Neubauer v Germany* were argued based on alleged breaches of relevant constitutional provisions which protect the environment, including for the benefit of future generations.

These successful cases also have in common the fact that they have been pursued by citizens against their own governments. Cases like *Sacchi* and *Duarte Agostinho*, which sought to hold foreign governments accountable for transnational climate impacts, have not been successful. These cases were both ruled inadmissible due to the claimants' failure to exhaust any local remedies available to them, though the European Court in *Duarte Agostinho* also refused to recognise any extraterritorial obligations.

The separation of powers between legislature and judiciary has also proved a hurdle in establishing liability for climate harms. In *ENvironnement JEUnesse* the Québec Court of Appeal held that only the legislature could determine whether Canada's emissions reduction policies were adequate. Similarly, in *Pandey v India* the court ruled that there could be no breach of the government's duties, since it could be assumed that the government was having regard to all relevant requirements in determining state policy on climate change. In *Sharma v Minister for Environment*, although the court at first instance ruled that the Minister owed a duty of care to children, on appeal this decision was overturned, with the appellate court ruling that the existence of such a duty was a matter for the legislature to implement, not to be inferred by the judiciary. These cases show the hesitancy of some courts to introduce 'new' obligations that have not been through a legislative process and the public scrutiny that entails.

Overall, the cases brought by children and young people show a cross-fertilisation of ideas and an improved success rate as arguments are refined. This upwards trajectory is also no doubt influenced by the reality that climate change is worsening, so that impacts that were once foreseeable risks have become likely occurrences, if not lived experience, for many young people. The recognition of this truth can be seen in many of the more recent judgments, in which courts acknowledge the urgency of tackling climate change and the very real prospect of irreversible harm if stronger action is not taken.

However, very few cases have successfully made out a violation of future generations' rights specifically. As noted earlier, *Future Generations v Colombia* upheld future generations' right to a healthy environment as found in the Colombian constitution. The judgments in *Neubauer* and *KlimaSeniorinnen* were strongly influenced by intergenerational equity and the need to avoid creating unjust burdens for future generations. The *Waratah Coal v Youth Verdict* judgment placed weight on the intergenerational aspects of First Nations' cultural rights. But all of these cases relied on children and young people to stand as litigants on behalf of future generations. We have yet to see cases explore the environmental rights of future generations in isolation from children's rights. As discussed in Chapter 4, using children as proxies for future generations is problematic. The following section considers whether a different approach might be possible.

7.5 LITIGATING THE ENVIRONMENTAL RIGHTS OF FUTURE GENERATIONS IN RELATION TO CLIMATE CHANGE

More than any other issue currently facing the international community, climate change threatens the environmental rights of future generations. It is therefore hardly surprising that the legal cases that have made the greatest advancements in protecting future generations' rights are within the field of climate litigation.

As discussed, future generations' environmental rights could be threatened by decisions relating to climate mitigation, including decarbonisation strategies and geoengineering projects, as well as by inadequate or inappropriate adaptation. As global warming worsens, there is no doubt that a failure to cut emissions will harm future generations, but turning this general statement of truth into a legal case in relation to a specific state or states' policies is still not straightforward. And cases relating to specific projects for decarbonisation, geoengineering or mitigation are far less common within human rights jurisprudence.

As the sample of cases discussed earlier illustrates, there has been considerable progress in recent years in recognising climate change as a human rights violation. To perhaps a lesser degree, there have also been developments in expanding the scope of states' duties under human rights laws to capture extraterritorial impacts of human rights and duties to prevent or protect against future harms. But the cases show that there are still barriers preventing full protection of future generations' environmental rights under international human rights law. This final section considers how those barriers might be overcome using the approach set out in Chapter 5.

As discussed in earlier chapters, a crucial first step is to recognise that future generations can be 'victims' of a human rights violation under the law. This is achieved by broadening the notion of victim status beyond circumstances of actual or imminent harm to include foreseeable harm. At the same time, we can expand the scope of states' duties to include these victims, following the three-step test explained in Chapter 5. Provided the state is exercising effective control over an act or omission, its human rights obligations will extend to the foreseeable consequences of that act/omission, bringing future generations within the ambit of the state's obligations.

This can be applied to any foreseeable harm associated with adaptation, geoengineering or decarbonisation projects that a state decides to pursue, since the related acts and decisions would be within its effective control. It would also extend to the foreseeable future harms of climate change itself, based on the effective control a state exercises over its climate policies, NDCs and emissions reduction strategies. The fact that other states might also be contributing to climate change would not change the fact that the state's actions have contributed to climate change in a way that was foreseeable.

To ensure that these cases can be brought to court, the rules of standing need to be changed to allow claims on behalf of future generations, as outlined in Chapter 5. The European Court of Human Rights has already shown the way in this respect, when it recognised standing for the organisation KlimaSeniorinnen to bring a case on behalf of its members who may be affected by climate change in the future. The next step would be to allow similar representative claims on behalf of future generations. This would address the problem explained in Chapter 4 of having to rely on children to act as proxies in litigation on behalf of future generations.

Protecting the environmental rights of future generations will be much easier if international laws and judicial bodies accept a more ecocentric interpretation of those rights. As cases like *Held v Montana* and *Future Generations v Colombia* show, having clearly recognised rights to a healthy or sustainable environment enables a stronger argument that climate laws and policies are in breach of human rights, since litigants do not need to show causal links to human harm as long as the threat of environmental harm is foreseeable.

There are other barriers to proving a violation which could be lowered using the approach explained in Chapter 5. Proving a violation will involve showing that the state has failed to uphold its duties and that the risk of harm to future generations is not justified. It is here that the theory of intergenerational equity becomes most useful, as it offers a principled way to balance the obligations that a state has to current as well as future generations, and the potentially conflicting obligations in relation to economic, social, cultural and environmental rights.

Proving causation may still be a challenge because of the cumulative causes of climate change, but as cases like *Neubauer*, *Waratah Coal*, *KlimaSeniorinnen* show, courts are more willing to hold states responsible for their contribution to global climate change based on shared responsibility and an understanding of climate science. An improved approach to causation – like the one suggested in Chapter 5 – will help with this even more. As that chapter explained, if we reconceptualise environmental rights in line with intergenerational equity, due diligence and the precautionary principle, we can similarly rethink what is required for causation. Causation then becomes not a question of whether a state's acts/omissions have caused the ultimate human rights impact, but rather whether its acts/omissions have addressed the risk of future harm. For example, a decision to allow geoengineering might be viewed as a breach of human rights if states have failed to adequately assess the potential environmental and human rights impacts and have not taken adequate measures to minimise the risk of future harm.

An inevitable issue when seeking to litigate future generations' environmental rights with respect to climate change will be how to balance their interests against those of current generations. The discussion at the beginning of this chapter illustrates this plainly – human rights issues arise from both climate change and the steps we take to address it, meaning there will always be a tension between the actions we take and the risks of not acting. In litigation, a respondent government might successfully argue that the negative impacts of building a wind farm, for example, are justified by the need to decarbonise and prevent future climate harms. To succeed with this argument however, the government would need to show that it has properly considered those future risks and done what it can to mitigate against them, including exploring alternative options.

The approach proposed in Chapter 5 does not mean that all climate litigation will be successful, but that is not its purpose. It is not intended to prevent any action that might have long-term consequences. Instead, it helps to achieve a fairer balance between current and future generations' rights by recognising that governments have duties to prevent foreseeable harm where possible, and that future generations' rights should only be limited where it is necessary and proportionate. The hope is that such an approach will help to prompt more serious action from states on tackling emissions and adapting to climate change, to help reduce the unjust burden that we leave for future generations.

8

Nuclear Energy

A Case Study of Future Environmental Rights Impacts of Nuclear Energy

8.1 INTRODUCTION

Nuclear energy facilities present a range of risks to future generations' rights, including some that raise new questions about how international human rights law might be applied. While the chances of a meltdown (e.g. Chernobyl) or natural disaster (e.g. Fukushima) are small, the gravity of the consequences would be severe and extremely long-lasting. The issue of how to safely store nuclear waste presents arguably less extreme risks, but still creates a latent threat to future generations and an ongoing burden for them in terms of managing waste materials which may remain hazardous for thousands of years.

Nuclear energy does offer benefits, however, most notably as an alternative to fossil fuels. Given the significant threats to human rights posed by climate change detailed in Chapter 7, these potential benefits will be relevant to any decision-making about acceptable risks for future generations. This final case study considers the human rights risks associated with nuclear energy, both in terms of the management of existing operations and the potential for new facilities to be opened. It shows how the approach outlined in Chapter 5 could be applied to safeguard the environmental rights of future generations.

8.2 CURRENT SCALE AND FUTURE ROLE OF NUCLEAR ENERGY

In the later decades of the twentieth century there was widespread uptake of nuclear energy in advanced economies, especially in Western Europe and North America. Currently there are around 440 nuclear reactors across thirty-two countries. Collectively, these contribute about 10 per cent of the world's total electricity supply.[1] Nuclear technology is also used for other purposes, including powering military submarines and medical diagnosis and treatment.

Nuclear power has long been considered a source of low-carbon energy and many of the scenarios which envisage a net-zero emissions future factor in a large contribution from nuclear energy.[2] For countries with an existing nuclear sector, nuclear energy plays a significant role in their plans to reduce emissions and meet their nationally determined contributions (NDCs) under the United Nations Framework Convention on Climate Change (UNFCCC). Article 4(2) of the Paris Agreement requires each member of the UNFCCC to declare its NDC, articulating its emissions reduction target and the steps it plans to take to get there.[3] Progress towards achieving the NDCs is assessed through a Global Stocktake, which is to occur every five years.[4]

In 2023, the first Global Stocktake took place at the twenty-eighth Conference of the Parties of the UNFCCC. As part of the Stocktake, the International Atomic Energy Agency (IAEA) provided an assessment of the role of nuclear energy in reducing emissions and meeting the overarching goals of the UNFCCC to keep global warming to 'well below 2 degrees Celsius'.[5] After analysing the NDCs and long-term strategies submitted by countries, the IAEA concluded that around 30 states have identified a role for nuclear energy in meeting their emissions reduction goals.[6] Of these, a little over twenty already have an active nuclear energy programme, while Turkey is actively constructing its first nuclear power plant. The Democratic People's Republic of Korea and Ghana include nuclear power in their NDCs, despite having no active nuclear energy programmes. A further four states (Australia,

1 World Nuclear Association, 'Nuclear Power in the World Today' (August 2023) <https://world-nuclear.org/information-library/current-and-future-generation/nuclear-power-in-the-world-today.aspx>; International Atomic Energy Agency, *Nuclear Energy and Climate Change: Questions and Answers on Progress, Challenges and Opportunities* (International Atomic Energy Agency, 2023) <www.iaea.org/publications/15503/nuclear-energy-and-climate-change-questions-and-answers-on-progress-challenges-and-opportunities>.

2 International Energy Agency, *World Energy Outlook 2022* (2022) <www.iea.org/reports/world-energy-outlook-2022>.

3 *Paris Agreement 2015* art 4(2).

4 Ibid art 14.

5 Ibid art 2(1)(a).

6 Argentina, Armenia, Australia, Canada, China, Colombia, Czech Republic, Democratic People's Republic of Korea, Finland, France, Ghana, Hungary, India, Islamic Republic of Iran, Japan, Mexico, Morocco, Netherlands, Russian Federation, Singapore, Slovakia, Slovenia, Sweden, Turkey, Ukraine, United Arab Emirates, United Kingdom, United States of America.

Colombia, Morocco, Singapore) refer to nuclear power in their longer-term strategies, but do not include it in their NDCs.

Yet, despite its role as a source of low-emissions electricity, there has been a decline in recent years in nuclear energy in many advanced economies, with plants closing and new investment decreasing.[7] The submission of the IAEA to the Global Stocktake mentioned earlier identifies thirteen states currently using or constructing nuclear power plants which do not include nuclear energy as part of their plans to cut emissions, suggesting a possible phasing out.[8] The IAEA has calculated that the average age of a nuclear plant in advanced economies is thirty-five years and that within the next five years as many as 25 per cent of reactors in these countries will close.[9] While most advanced economies may be moving away from nuclear, the number of reactors has increased in Asia and Eastern Europe, led mostly by new construction in China and Russia.[10] The World Nuclear Association reports that around thirty countries are considering, planning or starting nuclear power programmes. Several others are expanding their existing capacity.[11] It takes many years for a nuclear reactor to become operational, with lead times of between fifteen and twenty years not uncommon. At the other end of the lifespan, it can take a similar length of time to decommission a nuclear plant. The financial implications of either action are considerable. So too are the social and political considerations, particularly in relation to the location of plants, potential environmental and health risks, workplace relations and interaction with climate policy, among other factors. Decisions around whether to open, expand or decommission nuclear reactors are therefore complex and contentious, and raise significant issues with respect to the rights of future generations.

8.3 THE FUTURE HUMAN RIGHTS RISKS OF NUCLEAR ENERGY

Nuclear energy presents risks to future generations across the full life cycle of energy production. From the mining of uranium through its enrichment and

7 World Nuclear Association, *World Nuclear Performance Report 2022* (2022) 4 <www.world-nuclear.org/getmedia/9dafaf70-20c2-4c3f-ab80-f502488 3d9da/World-Nuclear-Performance-Report-2022.pdf.aspx>.

8 Bangladesh, Belarus, Belgium, Brazil, Bulgaria, Egypt, Germany, Republic of Korea, Pakistan, Romania, Switzerland, South Africa, Spain. International Atomic Energy Agency, *Nuclear Energy and Climate Change* (n 1) 6.

9 International Energy Agency, *Nuclear Power in a Clean Energy System* (2019) <www.iea.org/reports/nuclear-power-in-a-clean-energy-system>.

10 Ibid; World Nuclear Association, *Nuclear Power in the World Today* (n 7).

11 World Nuclear Association, 'Emerging Nuclear Energy Countries' (2023) <https://world-nuclear.org/information-library/country-profiles/others/emerging-nuclear-energy-countries.aspx>.

processing, to the operation of nuclear reactors and their eventual decommission, there are serious long-term risks which must be managed carefully to avoid potentially catastrophic consequences. These risks flow from the harmful nature of nuclear radiation and the extremely long lifespans of nuclear material, which can continue to emit hazardous radiation for up to a million years. The following discussion highlights some of the principal issues from a human rights perspective.

8.3.1 *Uranium Mining and Processing*

The health and environmental risks associated with nuclear energy begin with the mining of uranium. While uranium is safe while it remains underground, once extraction begins it presents a range of risks. The process of extracting uranium involves crushing the rock and adding it to water to create a slurry. The slurry is processed, usually with sulfuric acid, to extract the uranium in the form of uranium oxide (also known as yellowcake). The remaining slurry is transported to a tailings dam. The yellowcake is then further processed to produce enriched uranium ready for use. The enrichment process changes the composition of isotopes, going from a product that is mostly uranium-238 to one that is mostly the more fissile isotope uranium-235.

As a heavy metal, uranium has both radiological and chemical effects, and is most dangerous if ingested (e.g. if it enters water supplies). Exposure to uranium can lead to cancers, kidney disease, respiratory disorders, endocrine and neurological conditions, and fertility problems.[12] As it decays, uranium-238 produces a range of other damaging isotopes, with varying half-lives and different risk factors. One of its decay products is radium-226, which has a half-life of about 1,600 years and can be harmful with long exposure. Radium-226 decays into the more harmful gas, radon-222. Radon gas is invisible and easily distributed, and has been identified as the second highest cause of lung cancer after cigarette smoking.[13] Beyond radon-222, the decay chain of uranium-238 includes polonium and lead isotopes that emit alpha and beta radiation and can have different health impacts. Other uranium isotopes generate different decay products, which are also harmful in different ways. The use of sulfuric acid in the processing of uranium also releases other heavy metals which can be present in tailings dams, including zinc, copper, nickel and lead. These can have significant impacts on wildlife and human populations if they enter water supplies.

[12] *World Nuclear Waste Report* (2019) 47 <https://worldnuclearwastereport.org>.
[13] Ibid 48.

The health implications of uranium mining and processing are easily understood as human rights issues, particularly in relation to the right to the highest attainable standard of health. Where the right to a healthy environment is recognised then this right is also impacted. For workers, rights to safe working conditions are implicated, and there are also procedural concerns that might attach to decisions around whether to approve uranium mining. These sorts of issues are also relevant to the operation of nuclear power plants, considered in more detail in Section 8.3.2.

8.3.2 *Operation of Plants*

Health Impacts of Operation

The health impacts of nuclear power plants flow from the different forms of radiation which they emit. Radioactive materials used in nuclear power emit three kinds of radiation: alpha, beta and gamma. Alpha radiation is the most harmful, but also the most easily contained, since alpha particles cannot pass through paper or cloth barriers. Beta radiation can be contained by plastic or aluminium. Gamma is the least harmful form of radiation, but is still very hazardous, and needs thick concrete or lead to contain it.[14] Operating nuclear facilities produce both gases and liquids that can contain material that emits these kinds of radiation. The half-lives of these substances (the time it takes for half their volume to break down) can be thousands of years, meaning they can continue to emit radiation for potentially millions of years.

There has historically been some difficulty in ascertaining the exact level of risk from the operation of nuclear power plants. There have been some studies that have suggested an increase in cancer rates in surrounding areas, for example, increased rates of childhood leukaemia, but meta-analyses across all power plants are lacking, making it difficult to establish a causal relationship between nuclear energy and particular health impacts.[15] For example, apparent clusters of childhood leukaemia in areas surrounding nuclear power plants in the United Kingdom, France and Germany have been studied.[16] While the existence of a small number of clusters has been confirmed, scientists have not been able to attribute them to radiation exposure.[17]

[14] Ibid 45.

[15] Ibid 45, 50.

[16] Marek K Janiak, 'Epidemiological Evidence of Childhood Leukaemia around Nuclear Power Plants' (2014) 12(3) *Dose-Response* 349.

[17] D Laurier et al, 'Epidemiological Studies of Leukaemia in Children and Young Adults around Nuclear Facilities: A Critical Review' (2008) 132(2) *Radiation Protection Dosimetry* 182;

While there is understandable community concern about the cancer risks of nuclear power plants, scientific research has not established that normal levels of exposure in the surrounding area constitute a higher risk, though this work continues.[18]

We do know that the three forms of radiation are carcinogenic (causing cancer), mutagenic (causing changes to genetic material like DNA) and teratogenic (causing damage to embryos and foetuses). They also cause or contribute to cardiovascular disease, cataracts, stroke, mental illness and other health problems. The exact level of harm will depend on many factors, including the particular radioactive element (or radionuclide), the form in which it appears in the environment, its degree of solubility, and the individual's level of exposure.[19] People who have experienced a high dose of radiation to their whole bodies, even for a short period of time, can develop Acute Radiation Syndrome (ARS), which causes damage to the bone marrow, gastrointestinal and cardiovascular systems.[20] With the exception of milder doses where only the bone marrow is affected, ARS is usually fatal.

Risks of Meltdown

The most serious risks of nuclear energy occur in the event of an accident, natural disaster or other catastrophic incident. One of the earliest serious incidents was the partial meltdown at the Three Mile Island nuclear facility in Pennsylvania in 1979, but the most well-known examples are the Chernobyl meltdown in 1986 and the Fukushima meltdown of 2011.

The Chernobyl accident occurred when the power plant's Number Four reactor exploded. A surge of power during a test procedure led to a loss of coolant that caused the uranium fuel source to overheat. This then caused an explosion and meltdown of the reactor, in which the uranium core melted through the surrounding containment structures. A fire in the reactor core burned for several days, releasing radioactive particles into the air. The accident has been attributed to faulty design and inadequate training of operators.[21]

Daniel Cressey, 'Nuclear Power Plants Cleared of Leukaemia Link' [2011] *Nature* <www.nature.com/articles/news.2011.275>.

18 Antonello Russo et al, 'Incidence of Childhood Leukemia before and after Shut Down of Nuclear Power Plants in Germany in 2011: A Population-Based Register Study during 2004 to 2019' (2023) 152(5) *International Journal of Cancer* 913.

19 *World Nuclear Waste Report* (n 12) 45.

20 Centre for Disease Control, 'Acute Radiation Syndrome: A Fact Sheet for Physicians' (8 April 2022) <www.cdc.gov/nceh/radiation/emergencies/arsphysicianfactsheet.htm>.

21 International Atomic Energy Agency, 'Frequently Asked Chernobyl Questions' (Text, 7 November 2016) <www.iaea.org/newscenter/focus/chernobyl/faqs>.

Two plant workers died in the immediate explosion, while twenty-eight firefighters who had responded died from ARS in the months after the incident. The entire town of Pripyat, the nearest town to the power plant, was evacuated in the days immediately following the incident, and more people were evacuated from surrounding areas over subsequent months. In total, about 350,000 people were relocated.[22] The residents of Pripyat never returned and the town remains abandoned.

The explosion caused the emission of up to 100 different radioactive elements with varying degrees of toxicity. Contamination from Chernobyl affected an area of around 150,000 square kilometres, with fallout scattered across vast areas of the northern hemisphere. The effects of the accident also impacted food supplies and forced many changes to the way of life for people in the affected area.

The most notable health effects of the accident were experienced by those who were onsite at the time or responded in the days afterwards. In addition to the workers and firefighters who died, around 100 other responding workers experienced radiation sickness. There is strong evidence that children who were exposed to radiation from the Chernobyl incident were at a higher risk of developing thyroid cancer. This included children who lived in Belarus, Ukraine and parts of Russia who drank milk contaminated with radioactive iodine. In the period up to 2016, more than 11,000 thyroid cancer cases have been diagnosed in people who were children at the time of the incident.[23] There is some evidence of higher rates of leukaemia and cataracts among the people who worked on the recovery and clean-up operation that lasted several years and involved thousands of people.[24] People who did not experience direct health effects from radiation nonetheless experienced significant psychological, social and economic impacts, with high rates of anxiety and other mental illness reported.[25]

[22] United Nations Scientific Committee on the Effects of Atomic Radiation, 'Assessments of the Radiation Effects from the Chornobyl Nuclear Reactor Accident' (2008) <www.unscear.org/unscear/en/areas-of-work/chernobyl.html>.

[23] World Health Organization, *1986–2016: Chernobyl at 30: An Update* (25 April 2016) <https://cdn.who.int/media/docs/default-source/documents/publications/1986-2016-chernobyl-at-30476d83ab-985d-416f-b936–576b0c0ca17f.pdf?sfvrsn=434dbe67_1&download=true>; see also United Nations Scientific Committee on the Effects of Atomic Radiation (n 22); International Atomic Energy Agency, *Frequently Asked Chernobyl Questions* (n 21).

[24] World Health Organization (n 23) 2; United Nations Scientific Committee on the Effects of Atomic Radiation (n 22).

[25] Evelyn J Bromet, 'Emotional Consequences of Nuclear Power Plant Disasters' (2014) 106(2) *Health Physics* 206; EJ Bromet, JM Havenaar and LT Guey, 'A 25 Year Retrospective Review of the Psychological Consequences of the Chernobyl Accident' (2011) 23(4) *Clinical Oncology (Royal College of Radiologists (Great Britain))* 297; JM Havenaar et al, 'Long-Term

The meltdown of the reactors at the Fukushima Daiichi nuclear power plant in 2011 was caused by an earthquake (9.0-magnitude) and subsequent tsunami. The tsunami spread over the main island of Japan, Honshu, and caused massive destruction and the loss of more than 18,000 lives. When the earthquake struck, the plant's emergency systems kicked in and automatically shut down the reactors. Diesel generators began powering the cooling system for the reactor cores, but these were soon hit by the 14-metre wave of the tsunami, causing a loss of power. Without power to run the cooling system, the fuel cores in three of the reactors overheated, causing a meltdown. A series of explosions also occurred. Radioactive material was emitted into the atmosphere and the Pacific Ocean. An exclusion zone was established within a 20-kilometre radius around the nuclear plant and more than 150,000 people were relocated from their homes.

The Fukushima meltdown and explosions did not cause immediate serious health impacts. No deaths were reported in connection with radiation exposure and monitoring showed that radiation levels in residents and the environment were low.[26] In the years since, there has been no convincing evidence of increased rates of cancer or other conditions caused by radiation. There has, however, been strong evidence that the evacuation and uncertainty caused by the meltdown had a significant impact on the health and well-being of residents, particularly in the form of increased rates of anxiety.[27] Existing chronic diseases and other conditions were exacerbated by the stress of evacuation, while access to medical treatment was also detrimentally affected.[28]

After a decade of clean-up at the site, Japanese authorities approved the release of wastewater from the site into the Pacific Ocean. The water had been treated to reduce the risk of radioactivity, with the Japanese government claiming that levels of radioactivity had been brought down to below

Mental Health Effects of the Chernobyl Disaster: An Epidemiologic Survey in Two Former Soviet Regions' (1997) 154(11) *The American Journal of Psychiatry* 1605; World Health Organization (n 23) 2.

26 Dennis Normile, 'This Physician Has Studied the Fukushima Disaster for a Decade – and Found a Surprising Health Threat', *Science* (online, 4 March 2021) <www.science.org/content/article/physician-has-studied-fukushima-disaster-decade-and-found-surprising-health-threat>.

27 Chihiro Nakayama et al, 'Lingering Health-Related Anxiety about Radiation among Fukushima Residents as Correlated with Media Information Following the Accident at Fukushima Daiichi Nuclear Power Plant' (2019) 14(5) *PLoS ONE* e0217285.

28 Masaharu Tsubokura, 'Secondary Health Issues Associated with the Fukushima Daiichi Nuclear Accident, Based on the Experiences of Soma and Minamisoma Cities' (2018) 67(1) *Journal of the National Institute of Public Health* 71.

permitted levels and presented a negligible risk to the ocean.[29] However, the move attracted criticism, particularly as to whether there might be an accumulation of radioactive material in the marine food chain.[30] The IAEA is conducting ongoing scientific testing to determine the impact of the water release on the marine environment around Fukushima. To date, they have found that the approach to treating the water is consistent with relevant safety standards and that the 'discharges … would have a negligible impact to people and the environment'.[31]

While the meltdown in Fukushima resulted from a natural disaster, there was considerable criticism of both the Japanese government and the plant's operator, TEPCO, for not taking more preventative measures. In 2022 a court found a number of TEPCO executives liable for failing to prevent the meltdown, which the court said they should have been able to foresee.[32] Similar cases had also found the government liable for its role in failing to prevent the meltdown and evacuation, though some of these cases have been overturned on appeal.[33] A number of lawsuits remain pending in relation to the Fukushima disaster.

A nuclear meltdown raises many serious human rights issues. Not only are there the impacts of the meltdown itself, but there are also all the human

[29] Bianca Nogrady, 'Is Fukushima Wastewater Release Safe? What the Science Says' (2023) 618 (7967) *Nature* 894.

[30] Gavin Blair, 'Fukushima Fish with 180 Times Legal Limit of Radioactive Cesium Fuels Water Release Fears', *The Guardian* (online, 24 July 2023) <www.theguardian.com/environment/2023/jul/24/fukushima-fish-with-180-times-legal-limit-of-radioactive-cesium-fuels-water-release-fears>; 'Scientists Test Fukushima Fish after Nuclear Plant Water Release', *Reuters* (online, 19 October 2023) <www.reuters.com/world/asia-pacific/scientists-test-fukushima-fish-after-nuclear-plant-water-release-2023-10-19>; Nogrady (n 29); International Atomic Energy Agency, 'IAEA to Conduct First Extensive Sampling of Marine Environment near Fukushima Daiichi since Start of Treated Water Release' (Text, 10 October 2023) <www.iaea.org/newscenter/pressreleases/iaea-to-conduct-first-extensive-sampling-of-marine-environment-near-fukushima-daiichi-since-start-of-treated-water-release>.

[31] International Atomic Energy Agency, *IAEA Review of Safety Related Aspects of Handling ALPS-Treated Water at TEPCO's Fukushima Daiichi Nuclear Power Station: Report 1: First Review Mission to Japan after the Start of ALPS Treated Water Discharge (October 2023)* (January 2024) <www.iaea.org/sites/default/files/first_review_mission_report_after_start_of_alps_treated_water_discharge_oct_23.pdf>.

[32] Justin McCurry, 'Fukushima Nuclear Disaster: Ex-Bosses of Owner Tepco Ordered to Pay ¥13tn', *The Guardian* (online, 13 July 2022) <www.theguardian.com/world/2022/jul/13/fukushima-nuclear-disaster-ex-bosses-of-owner-tepco-ordered-to-pay-yen-13tn-tokyo-court>.

[33] Mari Yamaguchi, 'Japan Top Court: Government Not Responsible for Fukushima Disaster', *The Diplomat* (online, 17 June 2022) <https://thediplomat.com/2022/06/japan-top-court-government-not-responsible-for-fukushima-disaster>; 'Fukushima: Japan Court Finds Government Liable for Nuclear Disaster', *BBC News* (online, 17 March 2017) <www.bbc.com/news/world-asia-39303178>.

rights concerns that exist with any disaster or emergency situation. The physical effects of a nuclear meltdown have obvious implications for the rights to life and health. These are most acute for workers and first responders, but can also impact nearby residents, particularly if the meltdown is not managed appropriately. Access to food, water, shelter and health care will likely be disrupted as communities are relocated, transport networks interrupted and water and food supplies potentially contaminated.

Evacuation and relocation present a series of human rights concerns, beginning with an obvious interference with the right to freedom of movement and the right to enjoy private, family and home life. If the displacement is permanent, as was the case for the residents of Pripyat, the right to property is relevant. Furthermore, we know that displacement increases the risks of other human rights violations, including sexual and gender-based violence, separation of families, over-policing and other abuses of dignity.[34] Displacement also interrupts businesses, social and cultural connections, government and legal processes, and any number of activities that are important for a life of dignity and freedom.

The response of government agencies post-meltdown can also bring human rights issues. Rights to freedom of information and expression, assembly and protest, participation in decision-making, and access to justice are all likely to be affected, either as a necessary part of the response or an inadvertent consequence. While many of these rights can legitimately be limited or even derogated during times of national emergency, the scale, breadth and duration of human rights issues associated with a meltdown challenges the usual criteria of necessity and proportionality, making it difficult to articulate what a human rights-based approach to nuclear disaster should look like. And even though the likelihood of a meltdown comparable to Chernobyl or Fukushima is very low, the magnitude of harm that follows warrants consideration as a serious risk to human rights.

Human Rights Issues Surrounding Location and Governance of Power Plants

As outlined earlier, nuclear energy can pose risks to human and environmental health, even when operating according to best practice and in the absence of serious incidents. Because of these risks, communities have

[34] Bridget Lewis and Rowena Maguire, 'A Human Rights-Based Approach to Disaster Displacement in the Asia-Pacific' (2016) 6(2) *Asian Journal of International Law* 326; Megan Bradley and Roberta Cohen, 'Disasters and Displacement: Gaps in Protection' (2010) 1(1) *Journal of International Humanitarian Legal Studies* 95.

understandably expressed concerns when a new power plant has been proposed and anti-nuclear activism is an ongoing feature of political discourse in many nuclear countries. There are a number of human rights issues associated with the regulation of nuclear power and the conversations that surround it.

Any decision to open a new nuclear power plant must be made within the framework of procedural environmental rights, such as those found in the *Aarhus Convention*.[35] Given the potential for serious, long-term harm, the assessment of risk for a plant will need to be extremely rigorous, informed by the best possible science and widely open to scrutiny. The World Nuclear Waste Report notes that a lack of published data on the degree of hazard posed by different forms of waste and waste management activities has made it difficult to assess the causal connections between levels of exposure and health impacts.[36] Accuracy and transparency of risk assessments must be the goal if human rights are to be protected.

In countries with widespread nuclear power programmes, nuclear plants and waste facilities tend to be located in poorer areas, often populated by Indigenous people and other minority groups, raising issues of environmental justice.[37] This is in addition to the legacy of injustice from the extensive nuclear weapons testing that was conducted on the colonised lands of Indigenous peoples or the homes of ethnic minorities.[38] In order to ameliorate the risks of environmental injustice, public participation processes need to be effective and meaningful, and there must be adequate opportunities for decisions to be challenged through accessible legal processes. Where Indigenous peoples are concerned, the expectation for free, prior and informed consent as set out in the UN Declaration on the Rights of

[35] *Convention on Access to Information, Public Participation in Decision-Making and Access to Justice in Environmental Matters* 1998 ('*Aarhus Convention*').

[36] *World Nuclear Waste Report* (n 12) 45.

[37] Dean Kyne, 'Public Exposure to U.S. Commercial Nuclear Power Plants Induced Disasters' (2015) 6(3) *International Journal of Disaster Risk Science* 238; Kristin Shrader-Frechette, *What Will Work: Fighting Climate Change with Renewable Energy, Not Nuclear Power* (Oxford University Press, 2011); Matteo Cimellaro, '"We Need to Wake Up": Algonquin Leaders Sound Alarm over Planned Nuclear Waste Facility near Ottawa River', *Canada's National Observer* (online, 20 June 2023) <www.nationalobserver.com/2023/06/20/news/algonquin-leaders-sound-alarm-over-planned-nuclear-waste-facility>; 'Algonquin First Nations Oppose Nuclear Waste Site on Their Unceded Territory', *Madawaska Valley Current* (online, 21 June 2023) <madvalleycurrent.com/2023/06/21/algonquin-first-nations-oppose-nuclear-waste-site-on-their-unceded-territory>; Johanna Höffken and MV Ramana, 'Nuclear Power and Environmental Injustice' (2024) 13 *WIREs Energy and Environment* e498.

[38] Barbara Rose Johnston (ed), *Half-Lives and Half-Truths: Confronting the Radioactive Legacies of the Cold War*, vol 45 (School for Advanced Research Press, 2007).

Indigenous Peoples (UNDRIP) must be met. Anything less than this would constitute a violation of human rights.

Human rights are also relevant in relation to the different forms of public engagement and protest that can surround decision-making on nuclear energy. Even in countries where nuclear power has been used for decades, public opposition is common, particularly when new plants or waste facilities are proposed. There are many stories of nuclear protestors being arrested or persecuted (though these are most commonly in relation to nuclear weapons).[39] For example, in 1977 over one thousand people were arrested while protesting the construction of the Seabrook Nuclear Power Plant in New Hampshire.[40]

In Australia, a country without nuclear power plants, a proposal to purchase nuclear-powered submarines from the United States and the United Kingdom (the so-called 'AUKUS' agreement)[41] prompted widespread public criticism and protest. Under the deal, Australia will be obliged to dispose of all waste produced by the submarines, including disposing of spent fuel at the end of its lifespan. This has been a cause of public concern, including among First Nations people whose land was used for British nuclear testing in the past.[42] There have also been protests in the towns where submarines are proposed to be built and based.[43] To date, these protests have been largely peaceful and not subject to government crackdown. But they have also not resulted in any

[39] Herbert P Kitschelt, 'Political Opportunity Structures and Political Protest: Anti-Nuclear Movements in Four Democracies' (1986) 16(1) *British Journal of Political Science* 57; '145 Protesters Arrested at Nuclear Power Plant', *The New York Times* (online, 1984) <www.nytimes.com/1984/01/23/us/around-the-nation-145-protesters-arrested-at-nuclear-power-plant.html>; 'Catholic Activists Arrested for Anti-Nuclear Protest Outside UN', *National Catholic Reporter* <www.ncronline.org/news/catholic-activists-arrested-anti-nuclear-protest-outside-un>; 'Woman Arrested in SA Nuclear Protest', *SBS News* (online, 6 May 2016) <www.sbs.com.au/news/article/woman-arrested-in-sa-nuclear-protest/kojdfglgf>; 'U.S. Activists Arrested at European Air Bases Protesting U.S. Nuclear Weapons Stationed There', *Democracy Now!* <www.democracynow.org/2023/8/10/nuclear_protests_netherlands>.

[40] 'Over 2,000 Protesters Occupy the Seabrook Nuclear Power Plant Construction Site in New Hampshire' (26 January 2023) <www.history.com/this-day-in-history/protest-seabrook-nucear-power-plant>.

[41] Australian Parliament, 'Joint Leaders Statement on AUKUS' (13 March 2023) <www.pm.gov.au/media/joint-leaders-statement-aukus>.

[42] Sarah Collard and Donna Lu, 'Spectre of Maralinga Hangs over AUKUS Nuclear Waste for Indigenous Communities', *The Guardian* (online, 18 March 2023) <www.theguardian.com/australia-news/2023/mar/19/spectre-of-maralinga-hangs-over-aukus-nuclear-waste-for-indigenous-communities>.

[43] 'Thousands Rally in Australia against Potential Submarine Base', *Reuters* (online, 6 March 2023) <www.reuters.com/world/asia-pacific/thousands-rally-australia-against-potential-submarine-base-2023-05-06>.

change of policy from the Australian government, and there was no public consultation undertaken prior to the AUKUS agreement being struck.

The prospect of moving into nuclear power has also been a recent topic of debate in Australia, following a proposal from the opposition party to build seven nuclear power plants.[44] The proposed sites were mostly current or former coal-fired power plants or in areas that have relied on the fossil fuel industry. The proposal was made against the backdrop of ongoing disagreements about how Australia should meet its emissions reductions targets and how to ensure a just transition for communities and workers who have depended economically on fossil fuels. These questions have historically been fraught electoral issues in Australia, whose economy is heavily dependent on its natural resource wealth and which has been slow to adopt emissions reduction targets. In this context, it is easy to view the opposition party's proposal as being a short-term strategy to win votes in communities who are understandably nervous about the transition to renewable energy. The long-term economic and environmental implications of adopting nuclear power have received less attention in the political discussions.[45]

Human rights related to transparency and public participation in decision-making are not just important for their own sake. As Maria Rost Rublee has explained, they are also key to ensuring that new power plants have a social licence to operate.[46] Without a social licence, nuclear projects are likely to attract ongoing opposition through the various stages of planning and implementation. The consequences of failing to deal with nuclear matters openly

44 Liberal National Party, 'Australia's Energy Future (Media Release)', Peter Dutton MP (online, 19 June 2024) <www.peterdutton.com.au/dutton-littleproud-obrien-media-release-australias-energy-future>.

45 Michelle Grattan, 'Dutton Goes Nuclear, Proposing Seven Government-Owned Generators with the First Starting in 2030s', *The Conversation* (19 June 2024) <http://theconversation.com/dutton-goes-nuclear-proposing-seven-government-owned-generators-with-the-first-starting-in-2030s-232815>; Jane Norman, 'Dutton's Nuclear Power Plan Relies on Convincing the States and Managing Waste, but First There's an Election to Win', *ABC News* (online, 19 June 2024) <www.abc.net.au/news/2024-06-20/dutton-nuclear-plan-must-convince-states/103999636>; John Quiggin, 'Peter Dutton Has Promised to Solve Our Energy Problems – but His Nuclear Policy Still Leaves Australians in the Dark', *The Conversation* (19 June 2024) <http://theconversation.com/peter-dutton-has-promised-to-solve-our-energy-problems-but-his-nuclear-policy-still-leaves-australians-in-the-dark-232816>; Matt McDonald, 'Peter Dutton's Latest Salvo on Australia's Emissions Suggests Our Climate Wars Are Far from over', *The Conversation* (11 June 2024) <http://theconversation.com/peter-duttons-latest-salvo-on-australias-emissions-suggests-our-climate-wars-are-far-from-over-232144>.

46 Maria Rost Rublee, 'AUKUS: What to Do with Nuclear Waste?', *The Lowy Institute* (17 April 2023) <www.lowyinstitute.org/the-interpreter/aukus-what-do-nuclear-waste>; International Atomic Energy Agency, 'Maintaining a Social Licence' (18 January 2019) <www.iaea.org/resources/nuclear-communicators-toolbox/basics/trust/social-licence>.

and through democratic processes can be costly, and can even mean that planned projects fall over.[47] For Indigenous people, the importance of gaining free, prior and informed consent must not be overlooked. Failure to obtain consent before locating a waste dump on Indigenous lands, for instance, would constitute a violation of human rights law, damage relationships and cause ongoing harm to the people concerned.

For future generations, these issues are amplified by the fact that they have had no opportunity to contribute to decision-making about whether and where to open a new nuclear power plant, how to manage the waste that it produces, or how to regulate its ongoing operations. Instead, they are locked into the responsibility of dealing with a nuclear plant based on the decisions of generations before.

8.3.3 *Storage of Waste Material*

The safe storage of nuclear waste material is perhaps the biggest threat to human rights in the long term. As noted earlier, waste from mining and processing of uranium is hazardous and must be stored carefully. Waste is also produced from other nuclear activities and different aspects of the energy production process. It is the spent fuel from nuclear reactors which is the most dangerous, however, even though it only represents a small proportion of all waste produced from nuclear activities. It is classified as high-level nuclear waste. After being used in the fission process for three or four years, the fuel will be spent and need to be stored. It is highly dangerous material: 'an unshielded, freshly unloaded spent fuel element delivers a lethal dose at one-metre distance in less than one minute'.[48] And it can remain harmful for up to a million years.[49] The long-term storage of nuclear waste represents a significant problem for the rights of future generations. It also raises troubling questions of intergenerational equity.

The *Joint Convention on the Safety of Spent Fuel Management and on the Safety of Radioactive Waste Management*, adopted under the auspices of the IAEA and ratified by ninety states, is the key legal instrument regulating the management of nuclear waste. The Convention applies to waste from all civilian nuclear reactors and other civilian operations and sets out the general

47 MV Ramana, 'Why Technical Solutions Are Insufficient' in Achim Brunnengräber and Maria Rosaria Di Nucci (eds), *Conflicts, Participation and Acceptability in Nuclear Waste Governance: An International Comparison Volume III* (Springer Fachmedien, 2019) 21.

48 *World Nuclear Waste Report* (n 12) 52.

49 Christine Ro, 'The Staggering Timescales of Nuclear Waste Disposal', *Forbes* <www.forbes.com/sites/christinero/2019/11/26/the-staggering-timescales-of-nuclear-waste-disposal>.

obligation of states.[50] Article 4 says that: 'Each Contracting Party shall take the appropriate steps to ensure that at all stages of spent fuel management, individuals, society and the environment are adequately protected against radiological hazards'.[51] The Convention then sets out obligations in relation to the siting, design and operation of spent fuel management facilities, though these are also fairly general in nature, requiring states to take 'appropriate steps' to ensure that they provide effective protection.[52]

Current thinking suggests that the best approach to storage of spent nuclear fuel is deep geological disposal.[53] This involves burying the waste in repositories deep below the surface of the earth. Choosing an appropriate site for these repositories is challenging, since the site must be not just geologically suitable but also socially, culturally and politically feasible. The IAEA has published guidelines on deep disposal. It points to the need to identify sites that can be expected to remain suitable for thousands of years. In particular, it says that suitable sites must have a low groundwater content that has been stable for several thousand years. They must also be geologically stable, with no or only minor history of seismic, tectonic, climatic or other disruptive activity.[54]

The costs of locating an appropriate site and building the repository are enormous and have deterred some countries from pursuing long-term storage. Countries like Japan, with its geologically active landscape, do not have suitable sites for long-term waste disposal. Other countries have attempted to locate sites but have been unsuccessful in getting a project off the ground. Finland is the only country to have commenced construction of a deep geological repository for spent fuel, though there are a small number of deep repositories for intermediate- and low-level nuclear waste.[55] Of the thirty-two countries that currently have nuclear power plants, none have successfully implemented a real long-term solution to storing the resulting waste.[56]

50 *Joint Convention on the Safety of Spent Fuel Management and on the Safety of Radioactive Waste Management* 2014 3.

51 Ibid 4.

52 Ibid 11.

53 International Atomic Energy Agency, *Scientific and Technical Basis for the Geological Disposal of Radioactive Wastes* (No TRS413, International Atomic Energy Agency, 2003) <www.iaea.org/publications/6568/scientific-and-technical-basis-for-the-geological-disposal-of-radioactive-wastes>.

54 Ibid; International Atomic Energy Agency, *Geological Disposal Facilities for Radioactive Waste* (International Atomic Energy Agency, 2011) <www.iaea.org/publications/8535/geological-disposal-facilities-for-radioactive-waste>.

55 For example, the Waste Isolation Pilot Plant (WIPP) in New Mexico, which houses waste from nuclear weapons.

56 *World Nuclear Waste Report* (n 12).

The solution instead has been to focus on interim storage. In most cases this is in 'wet storage' – cooling ponds often located at the site of the nuclear power plant in which spent fuel is stored. These pools also bring risks, for both people and the environment. If the cooling in the pools fails, the water would evaporate within a few days and the fuel assemblies (the groups of fuel rods that contain the fissile material) can ignite. Another option for interim storage is in dry caskets but, while these are more stable, it is also a more costly system.

Another waste issue arises as nuclear plants come to the end of their lifetimes. This is particularly relevant for the countries that were early adopters of nuclear power. The International Energy Agency has reported that 'the nuclear fleet in advanced economies is 35 years old on average and many plants are nearing the end of their designed lifetimes. Given their age, plants are beginning to close, with 25% of existing nuclear capacity in advanced economies expected to be shut down by 2025'.[57] As more power plants are decommissioned, large volumes of additional waste will be produced, which will also need to be stored.[58]

With respect to the rights of future generations, the challenge of safe long-term storage of nuclear waste raises the obvious issue that the serious health risks presented by nuclear waste will remain a problem for generations to come. To protect future generations' rights, a technical solution is needed which can reliably and safely store waste for long periods of time. As noted earlier, geological disposal offers the best option for long-term storage, but progress on this has been lacking. Human rights protection requires commitments to fund research and development to develop technical solutions and ongoing funding commitments to ensure projects can be completed and sustainably managed. Governance structures also need to be put in place to ensure the long-term management to the highest safety standards. States with nuclear waste responsibilities need to view these as human rights obligations.

The timescale of nuclear waste raises the unique problem of ensuring that generations in the distant future will have the knowledge needed to continue the management of dangerous nuclear waste. Various projects over the years have looked at the question of how to communicate the risks of nuclear waste to people in the future, particularly in relation to hazardous waste that might be buried underground in remote locations. Some of the suggested solutions read like science fiction, from pictograms designed to overcome changes to

[57] International Energy Agency (n 59).

[58] *World Nuclear Waste Report* (n 12) 52–53; Robert Alvarez, *Spent Nuclear Fuel Pools in the U.S.: Reducing the Deadly Risks of Storage* (Institute for Policy Studies, 2011) <https://ips-dc.org/spent_nuclear_fuel_pools_in_the_us_reducing_the_deadly_risks_of_storage>.

language over the millennia, 'nuclear priesthoods' that will pass on the knowledge of nuclear dangers, to genetically modified cats which will change colour in the presence of radiation.[59] There is of course no way of guaranteeing that any of these methods would work. The surest way to avoid setting a deadly trap for unwitting future generations is to avoid nuclear waste in the first place. Failing that, a commitment to transparency and education may be the best means of ensuring that the knowledge of waste sites and their dangers is not lost.

8.4 LITIGATING NUCLEAR POWER

Bringing international human rights law to bear on the issue of nuclear power is a complex issue. While the risks of harm are serious, they are in most cases unlikely to eventuate, provided management of the nuclear fuel cycle is adequate. The vast time scale of nuclear waste products stretches beyond anything that international human rights law has dealt with. It cannot be accommodated within existing legal frameworks and is difficult to address using even the most future-focused revision of that system. A brief look at the few cases that have interrogated nuclear energy and waste helps to show the limitations of the law in this area.

8.4.1 *Nuclear Issues before the ICJ*

Concerns about the long-term risks of nuclear energy have not been litigated in the International Court of Justice, but that court has considered issues associated with the use of nuclear weapons, and the reasoning of the court in those cases does provide some indication of the concerns and the way they might be dealt with by international law.

The ICJ first considered nuclear activity in the *Nuclear Tests* cases involving New Zealand and Australia against France.[60] For some years, France had been conducting atmospheric testing of nuclear weapons in the South Pacific,

[59] Matt Langione, 'Will Art Save Our Descendants from Radioactive Waste?', *JSTOR Daily* (13 May 2015) <https://daily.jstor.org/can-we-use-art-to-warn-future-humans-about-radioactive-waste>; Donato Paolo Mancini, 'How Colour-Changing Cats Might Warn Future Humans of Radioactive Waste', *The Guardian* (online, 9 January 2017) <www.theguardian.com/environment/shortcuts/2017/jan/08/colour-changing-cats-warn-radioactive-waste-nuclear-plants-distant-descendants>; Mark Piesing, 'How to Build a Nuclear Warning for 10,000 Years' Time', *BBC Future* (online, 3 August 2020) <www.bbc.com/future/article/20200731-how-to-build-a-nuclear-warning-for-10000-years-time>.

[60] *Nuclear Tests Case (New Zealand v France) (Judgment)* [1974] ICJ Rep 457; *Nuclear Tests Case (Australia v France) (Judgment)* [1974] ICJ Rep 253.

and was proposing to continue this testing. New Zealand and Australia complained that the radioactive fallout from the testing would interfere with their territorial rights, as it would negatively impact both the environment and the human population. They also claimed that pollution of the ocean and limitations placed by France on aircraft and marine vessels travelling near the test sites would constitute a violation of their freedom to navigate the high seas.

For its part, France argued that there was no evidence that the fallout from the atmospheric tests was harmful to the people or territory of either New Zealand or Australia. In correspondence between the parties, France claimed that the 'concern which had been expressed as to the long-term effects of testing could not be based on anything other than conjecture'.[61]

In the short term, New Zealand and Australia sought preliminary measures, similar to an injunction, to prevent France from conducting any atmospheric tests until the merits of the dispute could be determined by the court. These were granted in 1973, with the court agreeing that there was sufficient evidence to support the allegations against France on a *prima facie* basis.[62]

Subsequent to the granting of the preliminary measures, France declared that it no longer intended to conduct atmospheric tests in the South Pacific. Such unilateral declarations are considered to be legally binding under international law and, since the court had no reason to expect France would not comply with its declaration, it concluded that the substance of the dispute between the parties no longer existed. In December 1974, the court ordered that the case be discontinued.[63]

In the years that followed the 1974 decision, France commenced an extensive programme of underground nuclear tests in the South Pacific. Once again, New Zealand and Australia expressed their concern about these tests, amid growing domestic and international condemnation. In 1995 New Zealand approached the ICJ once more, seeking a re-examination of the 1974 dispute based on France's new conduct.[64] This was a procedurally unusual move, since the ICJ Statute makes no provision for reopening a case in such a fashion. However, in its 1974 judgment, the court had observed in

[61] *Nuclear Tests Case (New Zealand v France) (Interim Protection Order)* [1973] ICJ Rep 135, [29].

[62] Ibid; *Nuclear Tests Case (Australia v France) (Interim Protection Order)* [1973] ICJ Rep 99.

[63] *Nuclear Tests Case (Australia v France) (Judgment)* (n 60); *Nuclear Tests Case (New Zealand v France) (Judgment)* (n 60).

[64] *Request for an Examination of the Situation in accordance with Paragraph 63 of the Court's Judgment of 20 December 1974 in the Nuclear Tests Case (New Zealand v France) (Order)* [1995] ICJ Rep 288.

paragraph 63 that 'if the basis of this Judgment were to be affected, the Applicant (New Zealand and Australia) could request an examination of the situation in accordance with the provisions of the Statute'.[65] Thus it had left the door open for New Zealand to return.

In these later proceedings, much turned on the fact that France had shifted away from atmospheric testing to underground testing, and whether these new tests 'affected' the 'basis' of the 1974 decision. New Zealand argued that the new testing did indeed affect the basis of the 1974 judgment and relied on paragraph 63 as the basis for jurisdiction in the 1995 claim. However, the majority of the court held that the basis of the 1974 ruling had been France's undertaking not to continue with atmospheric testing, and that only a return to atmospheric testing would affect the basis of the original decision.[66] Because the 1974 case had dealt 'exclusively' with atmospheric nuclear testing, the court ruled that it was unable to consider later claims in relation to underground testing and dismissed the case.[67]

The majority noted that the Order was 'without prejudice to the obligations of States to respect and protect the natural environment',[68] but otherwise did not engage with the substantive issues surrounding nuclear testing (atmospheric or underground). However, the dissenting opinion of Judge Weeramantry is very helpful in putting those issues in the context of contemporary environmental law.

He pointed to the fact that international environmental law had advanced significantly since 1974, arguing that such evolution of the law on its own affected the basis of the original decision. He also adopted a broader understanding of the basis for that decision, arguing that the focus of the case was on the potential harms caused by nuclear testing, not the specific mode of testing (atmospheric versus underground). As such, the resumption of testing by France could affect the basis of the 1974 decision, even though the tests were in a new form.[69]

[65] *Nuclear Tests Case (New Zealand v France) (Judgment)* (n 60) [63]; *Nuclear Tests Case (Australia v France) (Judgment)* (n 60) [60]; Matthew CR Craven, 'New Zealand's Request for an Examination of the Situation in Accordance with Paragraph 63 of the Court's Judgment of 20 December 1974 in the Nuclear Tests (New Zealand v France) Case, Order of 22 September 1995' (1996) 45(3) *The International and Comparative Law Quarterly* 725.

[66] *Request for an Examination of the Situation in Accordance with Paragraph 63 of the Court's Judgment of 20 December 1974 in the Nuclear Tests Case (New Zealand v France)* (n 64) [62].

[67] Ibid [63] and [68].

[68] Ibid [64].

[69] *Request for an Examination in the Nuclear Tests (New Zealand v France) (Dissenting Opinion of Judge Weeramantry)* (1992) 86 (International Court of Justice, July 1992) 317, 335.

Among his discussion, Judge Weeramantry identified the precautionary principle and the prohibition against transboundary harm among relevant environmental law principles that could constrain France's actions with respect to nuclear testing.[70] Together with the other dissenting judges, he explained that nuclear testing, while not *per se* unlawful, must be conducted in a way which respects these requirements of international environmental law. He paid particular attention to the principle of intergenerational equity, saying: 'The case before the Court raises, as no case ever before the Court has done, the principle of intergenerational equity – an important and rapidly developing principle of contemporary environmental law'.[71] Quoting Edith Brown Weiss, he pointed to New Zealand's right and responsibility to protect the interests of its unborn generations, explaining that these interests stood to be impacted by France's actions.

Weeramantry also reflected on the role of the ICJ in cases such as this one. He noted that, up until that point, the court had not had an opportunity to consider intergenerational equity, but that it was now presented with a 'pre-eminent opportunity to do so'.[72] He stated that the court 'must regard itself as a trustee of those (intergenerational rights) in the sense that a domestic court is a trustee of the interests of an infant to speak for itself … This consideration involved is too serious to be dismissed as lacking in importance merely because there is no precedent on which it rests'.[73]

Overall, however, the majority of the court in the *Nuclear Tests* cases did not take up the opportunity to engage meaningfully with the risks of nuclear testing or the potential dangers for future generations. It missed the opportunity to give guidance on how the developing rules of international environmental law might apply to nuclear activities.[74]

Not long after the *Nuclear Tests* re-examination, the court delivered its Advisory Opinion on the *Legality of the Threat or Use of Nuclear Weapons*.[75] The court had been asked by the United Nations General Assembly to provide advice on the question: 'Is the threat or use of nuclear weapons in any circumstance permitted under international law?'.[76] A previous request had

70 Ibid 342–347.
71 Ibid 341.
72 Ibid 342.
73 Ibid 641.
74 Craven (n 65) 734; Stephen M Tokarz, 'A Golden Opportunity Dismissed: The New Zealand v. France Nuclear Test Case' (1998) 26(4) *Denver Journal of International Law and Policy* 745.
75 *Legality of the Threat or Use of Nuclear Weapons (Advisory Opinion)* [1996] ICJ Rep 226.
76 UN General Assembly, 'Request for an Advisory Opinion from the International Court of Justice on the Legality of the Threat or Use of Nuclear Weapons (Resolution 49/75K)'

been made by the World Health Organization, but the court declined to answer on the basis that it was not a question that the WHO had authority to ask under the ICJ Statute and the UN Charter.[77] The General Assembly's request came on the back of considerable discussion on nuclear disarmament, a long-term programme of work that eventually led to the negotiation of the *Treaty on the Prohibition of Nuclear Weapons*, adopted in 2017.[78]

In its Advisory Opinion, the ICJ recognised the potentially catastrophic and long-term environmental consequences of the use of nuclear weapons, including the consequences for future generations:

> [T]he Court recognises that the environment is under threat and that the use of nuclear weapons could constitute a catastrophe for the environment. The Court also recognises that the environment is not an abstraction but represents the living space, the quality of life and very health of human beings, including unborn generations.[79]

The scale of the risk posed by nuclear weapons was recognised by the court, noting that:

> [t]he destructive power of nuclear weapons cannot be contained in either space or time. They have potential to destroy all civilisation and the entire ecosystem of the planet … Further, the use of nuclear weapons could be a serious danger to future generations. Ionising radiation has the potential to damage the future environment, food marine ecosystem, and to cause genetic effects and illnesses to future generations.[80]

In answering the question about the legality of their use, the court explained that there was no specific rule of international law governing nuclear weapons. Rather, states are subject to the same rules that apply to any other weapons, along with the general rules which regulate their actions. With respect to environmental impacts, some states had submitted that the prohibition against transboundary harm would preclude the use of nuclear weapons. The majority rejected this view, however. While the court confirmed that states have obligations to respect and protect the environment, including a duty not to cause transboundary environmental harm, these rules do not displace states' right of self-defence. Instead, they would form part of the

<https://documents-dds-ny.un.org/doc/UNDOC/GEN/N95/760/03/PDF/N9576003.pdf?OpenElement>.

77 *Legality of the Use by a State of Nuclear Weapons in Armed Conflict (Advisory Opinion)* [1996] ICJ Rep 66.

78 *Treaty on the Prohibition of Nuclear Weapons 2017.*

79 *Legality of the Threat or Use of Nuclear Weapons (Advisory Opinion)* (n 75) [29].

80 Ibid [55].

consideration of whether a use of force is 'necessary and proportionate in the pursuit of legitimate military objectives'.[81] And, while it recognised the significant impact that nuclear weapons would have on the environment left to future generations, the court stopped short of recognising the rights of future generations or relying on intergenerational equity.[82]

In his Dissenting Opinion, Judge Weeramantry was prepared to go further, however, and his comments on responsibility towards future generations in this context are worth including in detail. He said that:

> [t]he effects upon the eco-system extend, for practical purposes, beyond the limits of all foreseeable historical time ... At any level of discourse, it would be safe to pronounce that no generation is entitled, for whatever purpose, to inflict such damage on succeeding generations.
>
> The Court as the principal organ of the United Nations, empowered to state and apply international law with an authority matched by no other tribunal must, in its jurisprudence, pay due recognition to rights of future generations. If there is any tribunal that can recognise and protect their interests under the law, it is this Court.[83]

Judge Weeramantry also commented on the status of the rights of future generations:

> It is to be noted in this context that the rights of future generations have passed the stage when they were merely embryonic rights struggling for recognition. They have woven themselves into international law through major treaties, through juristic opinion and through general principles of law recognised by civilised nations.[84]

As can be seen, Judge Weeramantry reiterated his view that the ICJ, as the highest authority on international law, has an important role to play in safeguarding the interests of future generations. His comment that the rights of future generations are now recognised across international treaties and principles is a significant endorsement of their legal status, and no doubt helped to inspire many of the developments which have occurred in the three decades since (discussed in Chapter 3).

81 Ibid [30].

82 *Legality of the Threat or Use of Nuclear Weapons (Advisory Opinion)* (n 75); Malgosia Fitzmaurice, 'Intergenerational Equity Revisited' in Isabelle Buffard et al (eds), *International Law between Universalism and Fragmentation* (Brill Nijhoff, 2008) 224.

83 *Legality of the Threat or Use of Nuclear Weapons (Advisory Opinion) (Dissenting Opinion of Judge Weeramantry)* [1996] ICJ Rep 429, 454–455.

84 Ibid 456.

While the case focused on nuclear weapons and their particularly destructive capacity, concerns about the long-term impact of ionising radiation and other environmental consequences are arguably transferrable to nuclear energy generally. Indeed, where nuclear weapons will be governed by the *lex specialis* relating to the use of force and armed conflict, nuclear power does not have the same kind of special circumstances which might justify its negative effect. The ICJ's acknowledgement in both the *Nuclear Tests* cases and the *Nuclear Weapons Advisory Opinion* that states have obligations to respect and protect the environment suggests that a failure to contain the dangers of nuclear power might very well constitute a breach of international law.

8.4.2 *Nuclear Cases in the European Court of Human Rights*

A small number of cases have looked at the human rights implications of nuclear activities, principally in the European Court of Human Rights. In *LCB v UK*,[85] the applicant's father had worked with the British military during testing of nuclear weapons in the 1950s. He had been based on Christmas Island, off the Australian coast, which was the location of several tests. In 1970, when LCB was a young child, she was diagnosed with leukaemia. She argued that the British government failed to monitor her father's exposure to radiation and failed to warn her family about the risk that she might develop childhood cancer. She argued, among other things, that this constituted a violation of her right to life and to freedom from cruel or inhuman treatment under the European Convention on Human Rights.

After considering evidence about the nature of nuclear testing on Christmas Island and the connection between parental exposure to radiation and risks of cancer in their children, the Court found no violation of the European Convention had occurred. It concluded that there was inadequate evidence that LCB's father had in fact been exposed to harmful amounts of radiation during the tests and no reason for the government to have suspected he was unwell or otherwise affected.[86] Further, the Court also found that the evidence (both at the time of her diagnosis and at the time of the hearing) did not support a connection between parental exposure and childhood leukaemia. Without such a causal link, it could not find that the government had any reason to think she was at a higher risk of

85 *LCB v United Kingdom (Merits Judgment)* [1999] ECHR Eur Court HR, Application No 23413/94, Case No 14/1997/798/1001, 49.
86 Ibid 37.

developing leukaemia and therefore had no duty to warn her or monitor her health as a child.[87] The Court was also not convinced that, even if a causal link had been known, it would have made a material difference to her treatment after diagnosis or her health outcomes.[88]

The case illustrates the challenge of using existing human rights laws to address nuclear risks. In *LCB*, while the alleged nuclear exposure had already occurred, the Court was asked to consider the protective steps the state ought to have taken in response to a possible risk of harm. In that sense it is similar to a situation involving future harm. The claim failed because the Court felt the evidence did not establish a risk of harm from exposure that the state was required to address. Arguably, the Court should have taken a more precautionary interpretation of the evidence. Given that there was some uncertainty in the science around parental exposure to nuclear testing and childhood illness, the state should have erred on the side of caution, both in terms of protecting LCB's father and in relation to any ongoing risk to his children.

8.4.3 *Litigating the Future Rights Impacts of Nuclear Energy*

The discussion earlier shows that nuclear energy has potential to cause profound human rights impacts for future generations. The approach outlined in Chapter 5 could be employed by communities who are concerned about the future impacts of nuclear projects, but success in a legal proceeding is far from assured.

As a starting point, future generations would fall within the jurisdiction of the state based on its exercise of control over nuclear energy and the foreseeability that it will affect the environmental rights of future generations. As explained earlier, there are a range of potential impacts which could flow from uranium mining and the construction, operation, and decommissioning of a facility, in addition to the risks of a meltdown or other emergency situation. Given that these impacts would be foreseeable to the state, the revised approach to jurisdiction outlined in Chapter 5 would bring the future impacts within the state's jurisdiction, meaning that it owes duties to future generations with respect to the impacts on their environmental rights.

As explained in Chapter 5, the specific obligations placed on states will be based on the duties to respect, protect and fulfil human rights, informed by principles of intergenerational equity, due diligence and the precautionary principle. The duty to respect requires that states refrain from conduct that

87 Ibid 39.
88 Ibid 40.

limits the diversity or quality of the environment that future generations will enjoy. To comply with the duty to protect, states must regulate the actions of private entities to preserve diversity, quality and accessibility of the environment. Under the duty to fulfil, they must take positive steps to maintain, rehabilitate and improve environmental diversity, quality and access.

As with fracking, nuclear energy raises concerns at all three levels of duty. Arguably, the production of nuclear energy, especially uranium mining and storage of nuclear waste, limits the quality, diversity and accessibility of the environment in ways that cannot be avoided. The inherent dangers of radioactive substances mean that, while they can be managed and minimised, some degree of risk will always remain. The question of whether a state has breached its duties to respect, protect and fulfil environmental rights will come down to whether it has acted with due diligence in response to that risk.

Due diligence requires that the state conducts an adequate impact assessment. As discussed in Chapter 5, to properly consider the environmental rights of future generations, an impact assessment must consider the long-term human rights implications of the proposed activity. Conducting a meaningful assessment of the costs and benefits of a nuclear project, potentially across millennia, is a sizable challenge. While it is probably impossible to quantify the costs or benefits with any degree of specificity, an approach which gives meaningful consideration to the rights of future generations would at least ensure that those long-term costs are not discounted.

Where the due diligence process indicates that the project has potential to cause or contribute to serious or irreversible harm (as it would inevitably do with an assessment of nuclear activities, given the inherent dangers of radiation), the state would be obliged to put in place appropriate measures to prevent or mitigate those adverse impacts. To the extent that there is doubt about the future effects, the precautionary principle reinforces the need for a cautionary approach. Where a state fails to take these appropriate preventative or mitigation measures, it could be considered to be in breach of its human rights obligations.

The main problem in using human rights law to challenge a nuclear project would lie in showing that the state's regulation and administration of the project fall short of expectations. Provided a state adopts best practice in designing and constructing nuclear facilities and associated maintenance and safety protocols, it might be argued that it has done enough to mitigate potential risks. While the gravity of a potential meltdown or other incident would be high, these are uncommon and, provided the usual safeguards are in place, it might be difficult to show that a state has failed to take reasonable precautions. If an incident did occur then the precise causes would of course be the subject of detailed investigation, but it could be difficult to show a breach of obligation unless there was a clear failure in the planning, construction or management of the operations.

For future generations, however, the potential harm of nuclear energy is not limited to the possibility of a meltdown or other catastrophic event. Instead, the extremely long-lasting responsibility placed on future generations to manage nuclear waste should also be understood as an adverse impact. This suggests an alternative way of framing a claim under international human rights law, focused on the ways in which nuclear energy projects lock future generations into long-term costs.

These costs can be thought of in terms of limitations to future generations' quality, access and options with respect to the environment, to use the language of Edith Brown Weiss. The prospect of a nuclear meltdown would have obviously serious implications for environmental quality, access and options. However, as explained earlier, uranium mining, construction and decommissioning of power plants, and storage of nuclear waste all have impacts on the future use of land and quality of the environment. More broadly, the long-term commitment to a nuclear project represents an opportunity cost for future generations, whose options for other more nature-positive activities are restricted. So even without a meltdown, nuclear energy presents potential violations of future generations' environmental rights. This approach could be an especially potent argument in a country that did not already have a nuclear energy programme but was proposing to start one. By framing the human rights impact of nuclear power as a loss of equitable environmental access, quality and options based on the responsibility that is inherited by future generations, the harm being litigated would be less serious than a potential meltdown, but far more certain.

Even with this approach, there is still a potential argument that nuclear energy is a low-emissions alternative to fossil fuels and therefore is justified as a means of avoiding climate-related harms. The success of that argument would hinge on many factors, including the state's progress towards emissions reductions, the particular climate risks facing the country and how climate change is anticipated to affect future generations. These would need to be weighed up to determine whether the proposed nuclear activity was justifiable as an alternative source of energy. While the outcome of a specific claim is therefore unpredictable, an approach which recognises the environmental rights of future generations would at least ensure that their interests would be given due consideration.

8.5 CONCLUSION

This chapter has attempted to map the various ways in which nuclear energy impacts on the environmental rights of future generations. Some of these impacts are potentially catastrophic, albeit unlikely. Any nuclear activity entails a certain minimum level of harm for future generations, however, at least in terms of creating a responsibility to manage nuclear waste which has no foreseeable end date. This burden could be litigated perhaps, especially in countries proposing to embark on nuclear energy policy for the first time. Success is by no means guaranteed, and would likely come down to a judgment of whether procedural and preventative measures were adequate, or whether the future harms outweigh the potential benefits of the project. As a case study, nuclear energy highlights the critical importance of meaningful due diligence and long-term thinking when making decisions that will affect future generations.

9

Conclusion

Intergenerational injustice is plain to see in the way we have managed, exploited and neglected the environment. Equally plain are the critically important links between the environment and human rights. Yet, despite being fundamentally based on concerns of justice and equality, human rights law has failed to articulate the nature of our obligations with respect to future generations' human rights, or even to acknowledge that those rights exist. This book set out to find ways to improve international human rights law (IHRL) that might contribute to solving this problem.

The proposed reforms involve an explicit recognition of rights and duties held by future generations and states respectively, and a process for identifying when states' duties will be activated. That's the simple part. The specific expectations and applicable standards for states confronting a foreseeable impact on future generations' rights need to be determined in each set of circumstances with reference to a range of factors. The familiar tripartite duties to respect, protect and fulfil human rights need considerable fleshing out when the rights concerned belong to future generations.

I have suggested an approach that takes the respect/protect/fulfil framework and overlays it with principles of intergenerational equity, due diligence and the precautionary principle. The hope is that this gives a more detailed set of duties and standards that will guide states towards better long-term decision-making. Should they fail in this regard, it is also hoped the proposed reforms would enable advocates for future generations to take – and win – legal action to enforce the rights and duties.

To fine-tune the proposed model, it has been applied to three case studies in Chapters 6, 7 and 8. These examples illustrated the need for IHRL to focus more on due diligence and the precautionary principle in informing the content of states' obligations with respect to future generations. In this way, human rights duties can become more future-focused and address the need for prevention and protection across a longer timescale, not only when harm is imminent.

The revised approach does not mean that states cannot limit future generations' rights – it might be justifiable to impose a limitation, for example to prevent more serious or imminent harms caused by climate change. Weighing up these competing claims is difficult, however, and needs good evidence and reliable science. The case studies illustrate the critical importance of meaningful long-term risk assessment, investment in scientific research, and embedding the precautionary principle into all decision-making processes.

However, the case studies also show that human rights law cannot handle all issues – there are still some cases where harm might be considered too remote to give rise to obligations for current generations. Consequently, in addressing the risks of harm to future generations, we should not rely solely on human rights – we also need innovative political systems, strong democratic processes, and appropriate education to truly overcome short-termism.

The case studies were chosen not only because they represent long-term risks for the enjoyment of environmental rights, but also because they each raise subtly different challenges for the legal protection of those rights. Applying the new model from Chapter 5 to a variety of issues has helped to identify its strengths and weaknesses, as well as some of the limits of human rights-based approaches generally. To conclude the book, I want to draw a few final thoughts together, identify some lessons learned, and suggest some possible next steps.

Chapter 6 introduced the environmental rights issues associated with unconventional natural resource extraction, specifically fracking and deep seabed mining. Hydraulic fracturing (fracking) is a mining technique that presents environmental risks, both local and global, across different timescales. The practice is likely to cause localised impacts on soil, water and air, while the end product contributes to global greenhouse gas emissions. Some of these impacts may be short term, while others could take decades to materialise. There is also a degree of uncertainty about the long-term impacts of fracking, since it is a relatively new activity.

If uncertainty is a facet of fracking, it is the hallmark of deep seabed mining. Our scientific understanding of deep-sea ecosystems and biodiversity is so limited that we cannot know with any confidence how deep seabed mining might impact these areas. The example of deep seabed mining shows the importance of incorporating the precautionary principle into our understanding of environmental rights and duties, strengthening the idea of human rights due diligence so we can avoid severe, potentially irreversible environmental harm.

The case study of deep seabed mining also raises the question of who has rights with respect to areas beyond state jurisdiction. The deepest areas of the sea are considered the common heritage of mankind, but if these areas are damaged, who might be considered the 'victim'? Not only does this issue provoke questions about who has rights and duties with respect to the deep seabed, but also questions about how we define environmental rights, particularly whether we seek to protect the environment in its own right, or only because it benefits humans. The ecocentric/anthropocentric divide is an area where much more focused environmental rights scholarship is needed.

Chapter 7 then considered the many serious issues associated with climate change. The case study took a somewhat different approach, which was available because we already have such a wealth of climate litigation, including many cases that seek to uphold the rights of younger people and future generations. The analysis of these cases illustrated the challenge of addressing climate impacts which are foreseeable but not yet 'imminent'. The revised approach proposed in this book helps to overcome this challenge by reframing both victimhood and state duties so they apply to the foreseeable consequences of a state's exercise of control, even if those consequences may not materialise for some time.

Chapter 7 also shows that the potential environmental rights impacts of climate change flow not only from greenhouse gas emissions and global warming, but also from our responses to climate change, particularly decarbonisation, adaptation and geoengineering. Because environmental rights impacts can occur in all these different contexts, Chapter 7 raised important questions about how we should balance competing rights, especially the rights of current and future generations.

The final case study in Chapter 8 explored the various issues surrounding nuclear energy. The generation of nuclear power and the management of its by-products create environmental risks with a timescale that can span millennia. At their worst, these risks are catastrophic in the form of a meltdown or explosion, though very unlikely to materialise as long as proper procedures are in place. Less serious impacts are inevitable, however, and future generations will be unable to escape the responsibility to manage nuclear power plants and waste facilities to avoid environmental contamination.

The case study of nuclear energy shows the challenge of litigating on behalf of future generations when the environmental rights impacts are potentially severe but not likely to occur. States are unlikely to be in breach of their obligations as long as necessary due diligence and precautionary steps are taken to minimise the chance of a serious incident. However, as Chapter 8 explained, if we reframe our understanding of environmental rights and duties drawing on Edith Brown Weiss's theory of intergenerational equity, we can see that the potential impact for future generations isn't only the risk of a meltdown or other catastrophic incident. Future generations will experience a loss of access, quality and options with respect to the environment because they bear the responsibility to manage nuclear activities. For countries which are proposing to introduce nuclear power for the first time, litigation might challenge the burden which these decisions entail for future generations.

Together, the three case studies show how the revised approach to environmental rights might enable IHRL to better protect future generations from foreseeable harm. They also show that some of the questions that arise are not easy to answer, and not all litigation will be successful even with a modified approach to the theory and practice of environmental rights. But it is hoped that, even where the answer is not certain, we may nonetheless gain a clearer focus on what can be achieved through IHRL and what problems will require stronger action in other areas of endeavour.

Human rights law is an essential safeguard against abuse of governmental power and contains the ingredients for us all to enjoy lives of dignity, liberty and autonomy. Its design and delivery have prevented it from extending these same benefits to future generations, however, even where we can clearly see the consequences we are creating for them. Reforming human rights law so that future generations' rights are recognised and enforceable is a crucial first step towards addressing this problem.

Truly ensuring intergenerational justice will require a multifaceted reorganisation of many areas of law, politics, economics and social policy, however, and doing that will require some significant shifts in the way most communities think about their responsibilities towards future generations. There are valuable examples for this way of thinking among Indigenous peoples and many useful explanations from scholars of moral and political philosophy. It is hoped that this book might help share some of those ideas with new audiences, and that together we can move towards a more just, more ecologically reflexive, and more nature-positive version of international human rights law.

Bibliography

JOURNAL ARTICLES, NEWS STORIES AND REPORTS

'145 Protesters Arrested at Nuclear Power Plant', *The New York Times* (online, 1984) <www.nytimes.com/1984/01/23/us/around-the-nation-145-protesters-arrested-at-nuclear-power-plant.html>

Aceves, William J, 'Actio Popularis – The Class Action in International Law' [2003] (1) *The University of Chicago Legal Forum* 353

Aczel, Miriam R and Karen E Makuch, 'Human Rights and Fracking in England' (2018) 20(2) *Health and Human Rights* 31

Adelman, Sam, 'Geoengineering: Rights, Risks and Ethics' (2017) 8(1) *Journal of Human Rights and the Environment* 119

de Albuquerque, Catarina, *Report of the Special Rapporteur on the Human Right to Safe Drinking Water and Sanitation, Catarina de Albuquerque. Addendum: Mission to the United States* (UN, 2 August 2011) <https://digitallibrary.un.org/record/710677>

'Algonquin First Nations Oppose Nuclear Waste Site on Their Unceded Territory', *Madawaska Valley Current* (online, 21 June 2023) <https://madvalleycurrent.com/2023/06/21/algonquin-first-nations-oppose-nuclear-waste-site-on-their-unceded-territory/>

Allegretti, Aubrey and Sandra Laville, 'Liz Truss to Lift Fracking Ban "Despite Little Progress on Earthquake Risk"', *The Guardian* (online, 16 September 2022) <www.theguardian.com/environment/2022/sep/15/liz-truss-to-lift-fracking-ban-despite-little-progress-on-earthquake-risk>

Allen, Myles et al, *Intergovernmental Panel on Climate Change Special Report on Global Warming of 1.5oC: Summary for Policymakers* (Intergovernmental Panel on Climate Change, 6 October 2018)

Allen, Ted, 'The Philippine Children's Case: Recognizing Legal Standing for Future Generations Note' (1993) 6(3) *Georgetown International Environmental Law Review* 713

Alston, Philip, 'Making Space for New Human Rights: The Case of the Right to Development' (1988) 1 *Harvard Human Rights Yearbook* 3

'The Right to a Healthy Environment: Beyond Twentieth Century Conceptions of Rights' (2023) 117 *AJIL Unbound* 167

'The Shortcomings of a "Garfield the Cat" Approach to the Right to Development' (1985) 15(3) *California Western International Law Journal* 510

Alvarez, Ramón A et al, 'Greater Focus Needed on Methane Leakage from Natural Gas Infrastructure' (2012) 109(17) *Proceedings of the National Academy of Sciences* 6435

Alvarez, Robert, *Spent Nuclear Fuel Pools in the U.S.: Reducing the Deadly Risks of Storage* (Institute for Policy Studies, 2011) <https://ips-dc.org/spent_nuclear_fuel_pools_in_the_us_reducing_the_deadly_risks_of_storage/>

Anaya, S James, 'Report of the Special Rapporteur on the Situation of Human Rights and Fundamental Freedoms of Indigenous Peoples' Report to the United Nations Human Rights Council (UN Doc A/HRC/12/34/Add.5, 7 September 2009)

Arts, Karin, 'Children's Rights and Climate Change' in Claire Fenton-Glynn (ed), *Children's Rights and Sustainable Development* (Cambridge University Press, 2019) 216

Atapattu, Sumudu, *Human Rights Approaches to Climate Change* (Routledge, 2016)

'Intergenerational Equity and Children's Rights: The Role of Sustainable Development and Justice' in Claire Fenton-Glynn (ed), *Children's Rights and Sustainable Development: Interpreting the UNCRC for Future Generations* (Cambridge University Press, 2019) 167

Australian Parliament, 'Joint Leaders Statement on AUKUS' (13 March 2023) <www.pm.gov.au/media/joint-leaders-statement-aukus>

Australian Senate Standing Committees on Environment and Communication, *Oil and Gas Exploration and Production in the Beetaloo Basin: Second Progress Report* (April 2022)

Avila, Sofia, 'Environmental Justice and the Expanding Geography of Wind Power Conflicts' (2018) 13(3) *Sustainability Science* 599

Avila-Calero, Sofia, 'Contesting Energy Transitions: Wind Power and Conflicts in the Isthmus of Tehuantepec' (2017) 24(1) *Journal of Political Ecology* 992

Baldé, CP et al, *Global Transboundary E-Waste Flows Monitor 2022* (United Nations Institute for Training and Research, 2022) <www.scycle.info/global-transboundary-e-waste-flows-monitor-2022>

Barelli, Mauro, 'Free, Prior and Informed Consent in the Aftermath of the UN Declaration on the Rights of Indigenous Peoples: Developments and Challenges Ahead' (SSRN Scholarly Paper No 1991731, 25 January 2012) <https://papers.ssrn.com/abstract=1991731>

Baum, Seth D, 'The Great Downside Dilemma for Risky Emerging Technologies' (2014) 89(12) *Physica Scripta* 128004

Beckerman, Wilfred and Joanna Pasek, *Justice, Posterity and the Environment* (Oxford University Press, 2001)

Bell, Derek, 'Does Anthropogenic Climate Change Violate Human Rights?" (2011) 14(2) *Critical Review of International Social and Political Philosophy* 99

Bell-James, Justine and Briana Collins, 'Human Rights and Climate Change Litigation: Should Temporal Imminence Form Part of Positive Rights Obligations?' (2022) 13(1) *Journal of Human Rights and the Environment* 212

Bhattacharya, Pathikrit and Robert C Viesca, 'Fluid-Induced Aseismic Fault Slip Outpaces Pore-Fluid Migration' (2019) 364(6439) *Science* 464

Blair, Gavin, 'Fukushima Fish with 180 Times Legal Limit of Radioactive Cesium Fuels Water Release Fears', *The Guardian* (online, 24 July 2023)

<www.theguardian.com/environment/2023/jul/24/fukushima-fish-with-180-times-legal-limit-of-radioactive-cesium-fuels-water-release-fears>

Bodansky, Daniel, 'Climate Change and Human Rights: Unpacking the Issues' (2014) 38(3) *Georgia Journal of International & Comparative Law* 511

Deconstructing the Precautionary Principle (SSRN Scholarly Paper No ID 888831, 10 March 2006) <https://papers.ssrn.com/abstract=888831>

Bohaker, Heidi, '"Nindoodemag": The Significance of Algonquian Kinship Networks in the Eastern Great Lakes Region, 1600-1701' (2006) 63(1) *The William and Mary Quarterly* 23

Boston, Jonathan, *Governing for the Future: Designing Democratic Institutions for a Better Tomorrow* (Emerald Group Publishing, 2016)

Boucher, David, *The Limits of Ethics in International Relations: Natural Law, Natural Rights, and Human Rights in Transition* (Oxford University Press, 2011)

Boyd, David R, *A Safe Climate: Human Rights and Climate Change. Report of the UN Special Rapporteur on Human Rights and the Environment* (United Nations General Assembly, 15 July 2019)

Boyd, David and Stephanie Keene, *Essential Elements of Effective and Equitable Human Rights and Environmental Due Diligence Legislation* (Policy Brief No No 3, June 2022)

Boyd, David, John Knox and Marc Limon, *The Time Is Now: The Case for Universal Recognition of the Right to a Safe, Clean, Healthy and Sustainable Environment* (Universal Rights Group, February 2021)

Boyle, Alan, 'Human Rights and the Environment: Where Next?' (2012) 23(3) *European Journal of International Law* 613

Bradley, Megan and Roberta Cohen, 'Disasters and Displacement: Gaps in Protection' (2010) 1(1) *Journal of International Humanitarian Legal Studies* 95

Brent, Kerryn and Jeffrey McGee, 'The Regulation of Geoengineering: A Gathering Storm for International Climate Change Policy?' (2012) 24(4) *Air Quality and Climate Change* 22

Brent, Kerryn et al, 'International Law Poses Problems for Negative Emissions Research' (2018) 8(6) *Nature Climate Change* 451

Brent, Kerryn, Jeffrey McGee and Amy Maguire, 'Does the "No-Harm" Rule Have a Role in Preventing Transboundary Harm and Harm to the Global Atmospheric Commons from Geoengineering?' [2015] *Climate Law* 35

Brent, Kerryn, Jeffrey McGee and Jan McDonald, 'The Governance of Geoengineering: An Emerging Challenge for International and Domestic Legal Systems?' (2015-2016) 24(1) *Journal of Law, Information and Science* 1

Brigg, Morgan and Mary Graham, 'The Relevance of Aboriginal Political Concepts (5): Country, Place, and Territory', *ABC Religion & Ethics* (26 October 2020) <www.abc.net.au/religion/aboriginal-political-concepts-country-place-territory/12815608>

British Geological Survey, 'Shale Gas Extraction', *British Geological Survey* (2021) <www.bgs.ac.uk/geology-projects/shale-gas/shale-gas-extraction/>

Bromet, Evelyn J, 'Emotional Consequences of Nuclear Power Plant Disasters' (2014) 106(2) *Health Physics* 206

Bromet, EJ, JM Havenaar and LT Guey, 'A 25 Year Retrospective Review of the Psychological Consequences of the Chernobyl Accident' (2011) 23(4) *Clinical Oncology (Royal College of Radiologists (Great Britain))* 297

Brown, Bob, 'Harsher Anti-Protest Laws Targeting Environmentalists Are Putting Greed before Green', *The Guardian* (online, 27 June 2022) <www.theguardian.com/commentisfree/2022/jun/27/harsher-anti-protest-laws-targeting-environmentalists-are-putting-greed-before-green-bob-brown>

Brown Weiss, Edith, 'Climate Change, Intergenerational Equity, and International Law' (1985) 9 *Vermont Journal of Environmental Law* 615

'In Fairness to Future Generations and Sustainable Development' (1992) 8(1) *American University International Law Review* 19

In Fairness to Future Generations: International Law, Common Patrimony and Intergenerational Equity (Transnational, 1989)

'Our Rights and Obligations to Future Generations for the Environment' (1990) 84 *American Journal of International Law* 198

Brownlie, Ian, 'The Rights of Peoples in Modern International Law' (1985) 9 *Bulletin of the Australian Society of Legal Philosophy* 104

Burger, Michael and Maria Antonia Tigre, *Global Climate Litigation Report: 2023 Status Review* (Sabin Centre for Climate Change Law, Columbia Law School and United Nations Environment Programme, July 2023)

Burns, William CG, *The Paris Agreement and Climate Geoengineering Governance: The Need for a Human Rights-Based Component* (Report No 111, Centre for International Governance Innovation, October 2016)

Business and Human Rights Resource Centre, *Renewable Energy and Human Rights Benchmark: Key Findings from the Wind and Solar Sectors* (2021) <https://media.business-humanrights.org/media/documents/2021_Renewable_Energy_Benchmark_v5.pdf>

Buys, Elinor and Bridget Lewis, 'Environmental Protection through European and African Human Rights Frameworks' (2022) 26(6) *International Journal of Human Rights* 949

Calvin, Katherine et al, *Climate Change 2023: Synthesis Report. Contribution of Working Groups I, II and III to the Sixth Assessment Report of the Intergovernmental Panel on Climate Change (Summary for Policymakers)* (Intergovernmental Panel on Climate Change (IPCC), 25 July 2023)

Campbell-Duruflé, Christopher and Sumudu Anopama Atapattu, 'The Inter-American Court's Environment and Human Rights Advisory Opinion: Implications for International Climate Law' (2018) 8(3–4) *Climate Law* 321

Caney, Simon, 'Climate Change, Human Rights and Moral Thresholds' in Stephen Humphreys (ed), *Human Rights and Climate Change* (Cambridge University Press, 2009) 69

'Cosmopolitan Justice, Rights and Global Climate Change' (2006) 19(2) *Canadian Journal of Law and Jurisprudence* 255

'Human Rights, Climate Change and Discounting' (2008) 17(4) *Environmental Politics* 536

'Human Rights, Responsibilities and Climate Change' in Charles R Beitz and Robert E Goodin (eds), *Global Basic Rights* (Oxford University Press, 2009) 227

Carnegie Council for Ethics in International Affairs, 'Carnegie Climate Geoengineering Governance Initiative (C2G2)' <www.carnegiecouncil.org/programs/ccgg/index>

'Catholic Activists Arrested for Anti-Nuclear Protest outside UN', *National Catholic Reporter* <www.ncronline.org/news/catholic-activists-arrested-anti-nuclear-protest-outside-un>

Centre for Disease Control, 'Acute Radiation Syndrome: A Fact Sheet for Physicians' (8 April 2022) <www.cdc.gov/nceh/radiation/emergencies/arsphysicianfactsheet.htm>

Chandani, Achala, 'Distributive Justice and Sustainability as a Viable Foundation for the Future Climate Regime' (2007) 1(2) *Carbon & Climate Law Review* 12

Cimellaro, Matteo, '"We Need to Wake Up": Algonquin Leaders Sound Alarm over Planned Nuclear Waste Facility near Ottawa River', *Canada's National Observer* (online, 20 June 2023) <www.nationalobserver.com/2023/06/20/news/algonquin-leaders-sound-alarm-over-planned-nuclear-waste-facility>

Clarkson, Linda, Vern Morrissette and Gabriel Ragallet, *Our Responsibility to The Seventh Generation* (International Institute for Sustainable Development, 3 October 2001) <www.iisd.org/publications/our-responsibility-seventh-generation>

Collard, Sarah and Donna Lu, 'Spectre of Maralinga Hangs over AUKUS Nuclear Waste for Indigenous Communities', *The Guardian* (online, 18 March 2023) <www.theguardian.com/australia-news/2023/mar/19/spectre-of-maralinga-hangs-over-aukus-nuclear-waste-for-indigenous-communities>

Coward, Marcia W, Peter B Heller and Anna I Vellve Torras (tr.), 'Colombia's Constitution of 1991 with Amendments through 2005' <www.constituteproject.org/constitution/Colombia_2005.pdf>

Cox, Lisa, 'NT Government Deal to Buy Beetaloo Basin Gas from US Company Labelled "Carbon Bomb"', *Guardian Australia* (online, 23 April 2024) <www.theguardian.com/australia-news/2024/apr/23/nt-government-beetaloo-basin-gas-sale-plan-tamboran-resources>

Cranston, Maurice, *What Are Human Rights?* (Taplinger, 1973)

Craven, Matthew CR, 'New Zealand's Request for an Examination of the Situation in Accordance with Paragraph 63 of the Court's Judgment of 20 December 1974 in the Nuclear Tests (New Zealand v. France) Case, Order of 22 September 1995' (1996) 45(3) *The International and Comparative Law Quarterly* 725

Crawford, James (ed), *The Rights of Peoples* (Clarendon Press, 1988)

Cressey, Daniel, 'Nuclear Power Plants Cleared of Leukaemia Link' (2011) *Nature*. https://doi.org/10.1038/news.2011.275

Cullet, Philippe, 'Definition of an Environmental Right in a Human Rights Context' (1995) 13(1) *Netherlands Quarterly of Human Rights* 25

Cuyvers, Luc et al, *Deep Seabed Mining* (International Union for the Conservation of Nature, 2018)

Daly, Aoife, 'Climate Competence: Youth Climate Activism and Its Impact on International Human Rights Law' (2022) 22(2) *Human Rights Law Review* 1

'Intergenerational Rights Are Children's Rights: Upholding the Right to a Healthy Environment through the UN Convention on the Rights of the Child' (SSRN Scholarly Paper No 4141475, 20 June 2022) <https://papers.ssrn.com/abstract=4141475>

Daly, Erin and James May (eds), *Implementing Environmental Constitutionalism: Current Global Challenges* (Cambridge University Press, 2018)

D'Amato, Anthony, 'Do We Owe a Duty to Future Generations to Preserve the Global Environment?' (1990) 84(1) *American Journal of International Law* 190

Dauenhauer, Bernard, 'Response to Rawls' in Richard A Cohen and James L Marsh (eds), *Ricoeur as Another: The Ethics of Subjectivity* (State University of New York Press, 2002) 203

Denchak, Melissa, 'Fracking 101', *NRDC* (19 April 2019) <www.nrdc.org/stories/fracking-101>

Dick, Samantha, 'Gas Industry CEO Tells Senate Inquiry NT Fracking Cannot Happen without Traditional Owners' Consent', *ABC News* (online, 22 March 2022) <www.abc.net.au/news/2022-03-23/nt-beetaloo-basin-senate-inquiry-gas-industry-ceo/100929398>

Dierksmeier, Claus, 'John Rawls on the Rights of Future Generations' in Jörg Tremmel (ed), *Handbook of Intergenerational Justice* (Edward Elgar, 2006) 72

Doelle, Meinhard, 'Climate Change and Human Rights: The Role of the International Human Rights in Motivating States to Take Climate Change Seriously' (2004) 1(2) *Macquarie Journal of International and Comparative Environmental Law* 179

Donnelly, Jack, 'Human Rights as Natural Rights' (1982) 4(3) *Human Rights Quarterly* 391

'In Search of the Unicorn. The Jurisprudence and Politics of the Right to Development' (1985) 15(3) *California Western International Law Journal* 474

Drumbl, Mark A, 'Poverty, Wealth, and Obligation in International Environmental Law' (2001) 76(4) *Tulane Law Review* 843

Drupp, Moritz A et al, 'Discounting Disentangled' (2018) 10(4) *American Economic Journal: Economic Policy* 109

Dryzek, John S and Jonathan Pickering, *The Politics of the Anthropocene* (Oxford University Press, 2018)

Dworkin, Ronald, *Taking Rights Seriously* (Harvard University Press, 1978)

Elliot, Robert, 'The Rights of Future People' (1989) 6 *Journal of Applied Philosophy* 159

European Centre for Constitutional and Human Rights, 'Wind Park in Mexico: French Firm Disregards Indigenous Rights' (2020) <www.ecchr.eu/en/case/wind-park-in-mexico-french-firm-disregards-indigenous-rights/>

Feinberg, Joel, 'The Rights of Animals and Unborn Generations' in E Partridge (ed), *Responsibilities to Future Generations: Environmental Ethics* (Prometheus, 1981) 139

Finkel, Madelon L and Adam Law, 'The Rush to Drill for Natural Gas: A Five-Year Update' (2016) 106(10) *American Journal of Public Health* 1728

'The Rush to Drill for Natural Gas: A Public Health Cautionary Tale' (2011) 101(5) *American Journal of Public Health* 784

Finnis, John, *Natural Law and Natural Rights* (Oxford University Press, 2011)

Fisher, DE, *Australian Environmental Law: Norms, Principles & Rules* (Thomson Reuters, 3rd ed, 2014)

Fitzgerald, Roxanne, 'US Gas Company Tamboran Receives Approval to Expand Fracking in Beetaloo Basin', *ABC News* (online, 7 June 2024) <www.abc.net.au/news/2024-06-07/nt-government-approves-tamboran-exploratory-fracking-expansion/103950262>

Fitzmaurice, Malgosia, 'Intergenerational Equity Revisited' in Isabelle Buffard et al (eds), *International Law between Universalism and Fragmentation* (Brill Nijhoff, 2008) 195

Foulger, Gillian, 'Fracking Can Cause Earthquakes Tens of Kilometres Away – New Research', *The Conversation* <http://theconversation.com/fracking-can-cause-earthquakes-tens-of-kilometres-away-new-research-116539>

Friede, Stephanie, *Enticed by the Wind: A Case Study in the Social and Historical Context of Wind Energy Development in Southern Mexico* (Mexico Institute, 19 February 2016) <www.wilsoncenter.org/publication/enticed-the-wind-case-study-the-social-and-historical-context-wind-energy-development>

'Fukushima: Japan Court Finds Government Liable for Nuclear Disaster', *BBC News* (online, 17 March 2017) <www.bbc.com/news/world-asia-39303178>

Gardiner, Stephen, 'Why "Global Public Good" Is a Treacherous Term, Especially for Geoengineering' (2014) 123(2) *Climatic Change* 101

'A Call for a Global Constitutional Convention Focused on Future Generations' (2014) 28(03) *Ethics & International Affairs* 299

'Some Early Ethics of Geoengineering the Climate: A Commentary on the Values of the Royal Society Report' (2011) 20(2) *Environmental Values* 163

'Why Geoengineering Is Not a "Global Public Good", and Why It Is Ethically Misleading to Frame It as One' (2013) 121(3) *Climatic Change* 513

González-Ricoy, Iñigo, 'Constitutionalizing Intergenerational Provisions' in Iñigo González-Ricoy and Axel Gosseries (eds), *Institutions For Future Generations* (Oxford University Press, 2016) 170

González-Ricoy, Iñigo and Axel Gosseries, 'Designing Institutions for Future Generations' in Iñigo González-Ricoy and Axel Gosseries (eds), *Institutions For Future Generations* (Oxford University Press, 2016) 3

Goodman, Ryan, Henry J Steiner and Philip Alston, *International Human Rights in Context, 3rd Edition* (Oxford University Press, 2007)

Goolmeer, Teagan et al, 'Recognizing Culturally Significant Species and Indigenous-Led Management Is Key to Meeting International Biodiversity Obligations' (2022) 15(6) *Conservation Letters* e12899

Gosseries, Axel, 'On Future Generations' Future Rights' (2008) 16(4) *Journal of Political Philosophy* 446

Graham, Mary, *Custodial Navigator* 2013, <http://www.indigenoussovereigntyaustralia.com.au/wp-content/uploads/2013/06/CustodialNavigator.pdf>

'Some Thoughts about the Philosophical Underpinnings of Aboriginal Worldviews – AHR' [2008] (45) *Australian Humanities Review* <http://australianhumanitiesreview.org/2008/11/01/some-thoughts-about-the-philosophical-underpinnings-of-aboriginal-worldviews/>

Grantham Research Institute on Climate Change and the Environment, 'What Is Shale Gas, How Is It Extracted through Fracking and What Are Fracking's Impacts?' (11 February 2022) <www.lse.ac.uk/granthaminstitute/explainers/what-is-shale-gas-how-is-it-extracted-through-fracking-and-what-are-frackings-impacts/>

Grattan, Michelle, 'Dutton Goes Nuclear, Proposing Seven Government-Owned Generators with the First Starting in 2030s', *The Conversation* (19 June 2024) <http://theconversation.com/dutton-goes-nuclear-proposing-seven-government-owned-generators-with-the-first-starting-in-2030s-232815>

Guerrero, Marisa, 'New York State Codifies Fracking Ban in Budget', *NRDC* (3 April 2020) <www.nrdc.org/experts/marisa-guerrero/new-york-state-codifies-fracking-ban-budget>

Guterres, António, *Our Common Agenda: Report of the Secretary-General* (2021) <www.un.org/en/content/common-agenda-report/assets/pdf/Common_Agenda_Report_English.pdf>

To Think and Act for Future Generations (Policy Brief No 1, March 2023) <www.un.org/sites/un2.un.org/files/our-common-agenda-policy-brief-future-generations-en.pdf>

Haudenosaunee Confederacy, 'Values', *Haudenosaunee Confederacy* <www.haudenosauneeconfederacy.com/values/>

Hauter, Wenonah, 'Fracking's Inhumanity', *Center for Humans and Nature* (3 March 2017) <https://humansandnature.org/frackings-inhumanity/>

Havenaar, JM et al, 'Long-Term Mental Health Effects of the Chernobyl Disaster: An Epidemiologic Survey in Two Former Soviet Regions' (1997) 154(11) *The American Journal of Psychiatry* 1605

Herrera, Hector, '¿En qué va el fracking en el mundo?', *Razón Pública* (22 July 2019) <https://razonpublica.com/en-que-va-el-fracking-en-el-mundo/>

'The Legal Status of Fracking Worldwide: An Environmental Law and Human Rights Perspective', *Global Network for the Study of Human Rights and the Environment* (6 January 2020) <https://gnhre.org/human-rights/the-legal-status-of-fracking-worldwide-an-environmental-law-and-human-rights-perspective/>

Hill, Christopher B, Om P Yadav and Eakalak Khan, 'Hydraulic Fracturing Chemical Disclosure Policy and Data Analysis: Metrics and Trends in Transparency' (2021) 55(6) *Environmental Science & Technology* 3918

Hilson, Chris, 'The Visibility of Environmental Rights in the EU Legal Order: Eurolegalism in Action?' (2018) 25(11) *Journal of European Public Policy* 1589

Höffken, Johanna and MV Ramana, 'Nuclear Power and Environmental Injustice' (2024) 13 *WIREs Energy and Environment* e498

Holligan, Anna, 'Groningen Gas Fields – The Dutch Earthquake Zone', *BBC News* (online, 19 May 2013) <www.bbc.com/news/world-europe-22542982>

Horn, Laura, 'Climate Change and the Future Role of the Concept of the Common Concern of Humankind' (2015) 2 *Australian Journal of Environmental Law* 24

Horwitt, Dusty and Meilen Teklemichael, *Chemical Makers' Exemptions from Fracking Chemical Disclosure Rules* (Physicians for Social Responsibility, July 2022) < https://psr.org/wp-content/uploads/2022/07/chemical-makers-exemptions-from-fracking-chemical-disclosure-rules.pdf>

Hulme, Mike, *Can Science Fix Climate Change? A Case against Climate Engineering* (Polity Press, 2014)

'Increased Protections, Greater Certainty for Globally Iconic Queensland Section of Lake Eyre Basin', *Ministerial Media Statements* <https://statements.qld.gov.au/statements/99391>

Independent Expert Panel for the Legal Definition of Ecocide, 'Commentary and Core Text'

International Atomic Energy Agency, *Geological Disposal Facilities for Radioactive Waste* (International Atomic Energy Agency, 2011) <www.iaea.org/publications/8535/geological-disposal-facilities-for-radioactive-waste>

International Atomic Energy Agency, *IAEA Review of Safety Related Aspects of Handling ALPS-Treated Water at TEPCO's Fukushima Daiichi Nuclear Power Station: Report 1: First Review Mission to Japan after the Start of ALPS Treated Water Discharge (October 2023)* (January 2024) <www.iaea.org/sites/default/files/first_review_mission_report_after_start_of_alps_treated_water_discharge_oct_23.pdf>

International Atomic Energy Agency, *Nuclear Energy and Climate Change: Questions and Answers on Progress, Challenges and Opportunities* (2023) <www.iaea.org/publications/15503/nuclear-energy-and-climate-change-questions-and-answers-on-progress-challenges-and-opportunities>

International Atomic Energy Agency, *Scientific and Technical Basis for the Geological Disposal of Radioactive Wastes* (Report No TRS413, International Atomic Energy Agency, 2003) <www.iaea.org/publications/6568/scientific-and-technical-basis-for-the-geological-disposal-of-radioactive-wastes>

International Atomic Energy Agency, 'IAEA to Conduct First Extensive Sampling of Marine Environment Near Fukushima Daiichi Since Start of Treated Water Release' (Text, 10 October 2023) <www.iaea.org/newscenter/pressreleases/iaea-to-conduct-first-extensive-sampling-of-marine-environment-near-fukushima-daiichi-since-start-of-treated-water-release>

International Atomic Energy Agency, 'Maintaining a Social Licence' (18 January 2019) <www.iaea.org/resources/nuclear-communicators-toolbox/basics/trust/social-licence>

International Energy Agency, *Nuclear Power in a Clean Energy System* (2019) <www.iea.org/reports/nuclear-power-in-a-clean-energy-system>

International Energy Agency, *World Energy Outlook 2022* (2022) <www.iea.org/reports/world-energy-outlook-2022>

Janiak, Marek K, 'Epidemiological Evidence of Childhood Leukaemia around Nuclear Power Plants' (2014) 12(3) *Dose-Response* 349

Johnston, Barbara Rose (ed), *Half-Lives and Half-Truths: Confronting the Radioactive Legacies of the Cold War*, vol 45 (School for Advanced Research Press, 2007)

Joseph, Bob, 'What Is the Seventh Generation Principle?' <www.ictinc.ca/blog/seventh-generation-principle>

Kaplan, Thomas, 'Citing Health Risks, Cuomo Bans Fracking in New York State', *The New York Times* (17 December 2014) <www.nytimes.com/2014/12/18/nyregion/cuomo-to-ban-fracking-in-new-york-state-citing-health-risks.html>

Kerns, Thomas A and Kathleen Dean Moore (eds), *Bearing Witness: The Human Rights Case against Fracking and Climate Change* (Oregon State University Press, 2021)

Kerr, Richard A, 'Natural Gas from Shale Bursts onto the Scene' (2010) 328(5986) *Science* 1624

Kianga Judge, Sara, 'Starting Where You Are: First Nations Non-linear Storytelling', *The Australian Museum* (14 February 2024) <https://australian.museum/learn/first-nations/burra/non-linear-stories/australian.museum/learn/first-nations/burra/non-linear-stories/>

Kimmerer, Robin Wall, 'The Covenant of Reciprocity' in *The Wiley Blackwell Companion to Religion and Ecology* (Wiley, 2017) 368

Kingsford, RT and AJD Walburn, *Oil and Gas Exploration and Development in the Lake Eyre Basin: Distribution and Consequences for Rivers and Wetlands, including the Coongie Lakes Ramsar Site* (CSIRO Research Paper, 4 October 2022)

Kitschelt, Herbert P, 'Political Opportunity Structures and Political Protest: Anti-Nuclear Movements in Four Democracies' (1986) 16(1) *British Journal of Political Science* 57

Knox, John, *Climate Change and Human Rights* (Annual Report of the Special Rapporteur on Human Rights and the Environment to Human Rights Council, 3 March 2016)

'Climate Change and Human Rights Law' (2009) 50 *Virginia Journal of International Law* 163

Framework Principles on Human Rights and the Environment (UN Doc A/HRC/37/59 (annex), 2018)

'Greening Human Rights', *openDemocracy* (14 July 2015) <www.opendemocracy.net/en/openglobalrights-openpage/greening-human-rights/>

'Linking Human Rights and Climate Change at the United Nations' (2009) 33 *Harvard Environmental Law Review* 477

Knox, John and David Boyd, *Report of the Special Rapporteur on the Issue of Human Rights Obligations Relating to the Enjoyment of a Safe, Clean, Healthy and Sustainable Environment* (United Nations General Assembly, 19 July 2018)

'Constructing the Human Right to a Healthy Environment' (2020) 16(1) *Annual Review of Law and Social Science* 79

Kyne, Dean, 'Public Exposure to U.S. Commercial Nuclear Power Plants Induced Disasters' (2015) 6(3) *International Journal of Disaster Risk Science* 238

Langione, Matt, 'Will Art Save Our Descendants from Radioactive Waste?', *JSTOR Daily* (13 May 2015) <https://daily.jstor.org/can-we-use-art-to-warn-future-humans-about-radioactive-waste/>

Laurier, D et al, 'Epidemiological Studies of Leukaemia in Children and Young Adults around Nuclear Facilities: A Critical Review' (2008) 132(2) *Radiation Protection Dosimetry* 182

Laville, Sandra, 'Extinction Rebellion Founder Cleared over King's College Protest', *The Guardian* (online, 9 May 2019) <www.theguardian.com/environment/2019/may/09/extinction-rebellion-founder-cleared-over-kings-college-protest>

Lawrence, Peter, *Justice for Future Generations: Climate Change and International Law* (Edward Elgar, 2015)

'Justifying Representation of Future Generations and Nature: Contradictory or Mutually Supporting Values?' (2022) 11(3) *Transnational Environmental Law* 553

Lee, Patrick G, 'EPA Concludes Fracking a Threat to U.S. Water Supplies', *ProPublica* (online, 14 December 2016) <www.propublica.org/article/epa-concludes-fracking-a-threat-to-u.s.-water-supplies>

Legg, Andrew, 'Different Approaches to Deference in International Human Rights Law' in *The Margin of Appreciation in International Human Rights Law* (Oxford University Press, 2012)

Lehmann, Rosa and Anne Tittor, 'Contested Renewable Energy Projects in Latin America: Bridging Frameworks of Justice to Understand "Triple Inequalities of Decarbonisation Policies"' [2021] *Journal of Environmental Policy & Planning* 1

Lewis, Bridget, 'Balancing Human Rights in Climate Policies' in Ottavio Quirico and Mouloud Boumghar (eds), *Climate Change and Human Rights* (Routledge, 2015) 39

'Children's Human Rights-Based Climate Litigation at the Frontiers of Environmental and Children's Rights' (2021) 39(2) *Nordic Journal of Human Rights* 180

Environmental Human Rights and Climate Change: Current Status and Future Prospects (Springer, 2018)

'Environmental Rights or a Right to the Environment? Exploring the Nexus between Human Rights and Environmental Protection' (2012) 8 *Macquarie Journal of Comparative and International Environmental Law* 36

'Human Rights' in Gerry Nagtzaam, Katie O'Bryan and Mark Beaufoy (eds), *Legal Pathways to Deep Decarbonisation in Australia* (LexisNexis, 2023)

'Human Rights Duties towards Future Generations and the Potential for Achieving Climate Justice' (2016) 34(3) *Netherlands Quarterly of Human Rights* 206

'Protecting Environmental Human Rights for Future Generations' in Walter F Baber and James May (eds), *Environmental Human Rights in the Anthropocene: Concepts, Contexts, and Challenges* (Cambridge University Press, 2023)

'Sacchi et al v Argentina, Brazil, France, Germany and Turkey (United Nations Committee on the Rights of the Child)' (2022) 36(9/10) *Australian Environment Review* 201

'The Rights of Future Generations within the Post-Paris Climate Regime' (2018) 7(1) *Transnational Environmental Law* 69

Lewis, Bridget and Rowena Maguire, 'A Human Rights-Based Approach to Disaster Displacement in the Asia-Pacific' (2016) 6(2) *Asian Journal of International Law* 326

Liberal National Party, 'Australia's Energy Future (Media Release)', *Peter Dutton MP* (online, 19 June 2024) <www.peterdutton.com.au/dutton-littleproud-obrien-media-release-australias-energy-future/>

Linehan, Jan and Peter Lawrence, *Giving Future Generations a Voice: Normative Frameworks, Institutions and Practice* (Edward Elgar, 2021)

Lock the Gate, 'Lake Eyre Basin Protections Must Be Priority for Soon to Be Premier Steven Miles' (12 December 2023) <www.lockthegate.org.au/lake_eyre_basin_protections_must_be_priority_for_soon_to_be_premier_steven_miles>

MacKenzie, Michael K, *Institutional Design and Sources of Short-Termism* (Oxford University Press, 2016)

Macklin, Ruth, 'Can Future Generations Correctly Be Said to Have Rights?' in E Partridge (ed), *Responsibilities to Future Generations: Environmental Ethics* (Prometheus, 1981) 151

Maloney, Michelle, 'Contaminated Life: The True Cost and Human Rights Impacts of Unconventional Gas Mining in Australia', *Center for Humans and Nature* (20 April 2018) <https://humansandnature.org/contaminated-life-the-true-cost-and-human-rights-impacts-of-unconventional-gas-mining-in-australia/>

Mancini, Donato Paolo, 'How Colour-Changing Cats Might Warn Future Humans of Radioactive Waste', *The Guardian* (online, 9 January 2017) <www.theguardian

.com/environment/shortcuts/2017/jan/08/colour-changing-cats-warn-radioactive-waste-nuclear-plants-distant-descendants>

Martin, Neil, 'Seven Things You Need to Know about Lithium-Ion Battery Safety', *UNSW Sites* (20 March 2023) <www.unsw.edu.au/newsroom/news/2023/03/seven-things-you-need-to-know-about-lithium-ion-battery-safety>

Maule, Alexis L et al, 'Disclosure of Hydraulic Fracturing Fluid Chemical Additives: Analysis of Regulations' (2013) 23(1) *New Solutions* 167

May, James and Erin Daly, *Global Environmental Constitutionalism* (Cambridge University Press, 2015)

McCorquodale, Robert and Penelope Simons, 'Responsibility beyond Borders: State Responsibility for Extraterritorial Violations by Corporations of International Human Rights Law' (2007) 70(4) *The Modern Law Review* 598

McCosker, Maddelin and Victoria Pengilly, 'Origin Energy Backs Away from Outback Fracking Plans but Farmers Say Fight Isn't Over', *ABC News* (online, 22 September 2022) <www.abc.net.au/news/2022-09-23/origin-energy-backs-out-gas-exploration-channel-country-fracking/101456466>

McCurry, Justin, 'Fukushima Nuclear Disaster: Ex-Bosses of Owner Tepco Ordered to Pay ¥13tn', *The Guardian* (online, 13 July 2022) <www.theguardian.com/world/2022/jul/13/fukushima-nuclear-disaster-ex-bosses-of-owner-tepco-ordered-to-pay-yen-13tn-tokyo-court>

McDonald, Matt, 'Peter Dutton's Latest Salvo on Australia's Emissions Suggests Our Climate Wars Are Far from Over', *The Conversation* (11 June 2024) <http://theconversation.com/peter-duttons-latest-salvo-on-australias-emissions-suggests-our-climate-wars-are-far-from-over-232144>

McGee, Jeffrey, Kerryn Brent and Wil Burns, 'Geoengineering the Oceans: An Emerging Frontier in International Climate Change Governance' (2018) 10(1) *Australian Journal of Maritime & Ocean Affairs* 67

McIlwraith, Phoebe, 'Phoebe McIlwraith on Dancing with Country' [2022] *Groundswell* <www.groundswellgiving.org/journal/phoebe-mcilwraith>

McKinnon, Catriona, 'Sleepwalking into Lock-in? Avoiding Wrongs to Future People in the Governance of Solar Radiation Management Research' [2018] *Environmental Politics* 1

Morrow, David, 'International Governance of Climate Engineering: A Survey of Reports on Climate Engineering, 2009–2015' [2017] *SSRN Electronic Journal* <www.ssrn.com/abstract=2982392>

Morrow, David R, 'Why Geoengineering Is a Public Good, Even If It Is Bad' (2014) 123(2) *Climatic Change* 95

Müllerová, Hana, 'Environment Playing Short-Handed: Margin of Appreciation in Environmental Jurisprudence of the European Court of Human Rights' (2015) 24(1) *Review of European, Comparative & International Environmental Law* 83

Myers, Tom, 'Potential Contaminant Pathways from Hydraulically Fractured Shale to Aquifers' (2012) 50(6) *Groundwater* 872

Nakayama, Chihiro et al, 'Lingering Health-Related Anxiety about Radiation among Fukushima Residents as Correlated with Media Information Following the Accident at Fukushima Daiichi Nuclear Power Plant' (2019) 14(5) *PLoS ONE* e0217285

Neidjie, Bill, *Story about Feeling* (Magabala Books, 1989)

Niner, Holly J et al, 'Deep-Sea Mining with No Net Loss of Biodiversity – An Impossible Aim' (2018) 5 *Frontiers in Marine Science* 1

Nogrady, Bianca, 'Is Fukushima Wastewater Release Safe? What the Science Says' (2023) 618(7967) *Nature* 894

'Cobalt Is Critical to the Renewable Energy Transition. How Can We Minimize Its Social and Environmental Cost?', *Ensia* (14 May 2020) <https://ensia.com/features/cobalt-sustainability-batteries/>

Nolan, Aoife, 'Children's Rights and Climate Change at the UN Committee on the Rights of the Child: Pragmatism and Principle in Sacchi v Argentina', *EJIL: Talk!* (20 October 2021) <www.ejiltalk.org/childrens-rights-and-climate-change-at-the-un-committee-on-the-rights-of-the-child-pragmatism-and-principle-in-sacchi-v-argentina/>

'Children and Future Generations Rights before the Courts: The Vexed Question of Definitions' (2024) 13(3) *Transnational Environmental Law* (in press)

Norman, Jane, 'Dutton's Nuclear Power Plan Relies on Convincing the States and Managing Waste, but First There's an Election to Win', *ABC News* (online, 19 June 2024) <www.abc.net.au/news/2024-06-20/dutton-nuclear-plan-must-convince-states/103999636>

Normile, Dennis, 'This Physician Has Studied the Fukushima Disaster for a Decade – and Found a Surprising Health Threat', *Science* (online, 4 March 2021) <www.science.org/content/article/physician-has-studied-fukushima-disaster-decade-and-found-surprising-health-threat>

Nowak, Manfred, *U.N. Covenant on Civil and Political Rights: CCPR Commentary* (N.P. Engel, 2nd rev. ed., 2005)

Office of the Commissioner for Fundamental Rights of Hungary, *The Role of the Ombudsman for Future Generations* <www.ajbh.hu/web/ajbh-en/the-role-of-the-ombudsman>

Office of the High Commissioner for Human Rights, *Report of the Office of the United Nations High Commissioner for Human Rights on the Relationship between Climate Change and Human Rights* (15 January 2009)

'Over 2,000 Protesters Occupy the Seabrook Nuclear Power Plant Construction Site in New Hampshire' (26 January 2023) <www.history.com/this-day-in-history/protest-seabrook-nucear-power-plant>

'Pacific Island Countries Make Historic Grassroots Declaration on Climate Change Emergency in Kioa, Fiji' <www.caritas.org.au/news/media-releases/pacific-island-countries-make-historic-grassroots-declaration-on-climate-change-emergency-in-kioa-fiji/>

Page, Edward, *Climate Change, Justice and Future Generations* (Edward Elgar, 2006)

Papantoniou, Angeliki, 'Advisory Opinion on the Environment and Human Rights' (2018) 112(3) *American Society of International Law* 460

Parajuly, Keshav et al, *Future E-Waste Scenarios* (StEP, UNU ViE-SCYCLE & UNEP IETC, 2019)

Parfit, Derek, *Reasons and Persons* (Clarendon, 1984)

Parker, Andy and Peter J Irvine, 'The Risk of Termination Shock from Solar Geoengineering' (2018) 6(3) *Earth's Future* 456

'Parliamentary Friends for Future Generations – EveryGen' (27 March 2023) <www.everygen.online/2023/03/27/parliamentary-friends-for-future-generations/>
Peace Brigades International, *Wind Farms and Concerns about Human Rights Violations in Oaxaca* (14 March 2014) <https://pbi-mexico.org/fileadmin/user_files/projects/mexico/files/PBI_Publications/1403BriefingWindFarmsPBI.pdf>
Pedersen, Ole, 'Climate Change and Human Rights: Amicable or Arrested Development?' (2010) 1(2) *Journal of Human Rights and the Environment* 236
Pedersen, Ole W, 'European Court of Human Rights and Environmental Rights' in *Elgar Encyclopedia of Environmental Law* (Edward Elgar, 2019) 463
'The European Convention of Human Rights and Climate Change – Finally!', *EJIL: Talk!* (22 September 2020) <www.ejiltalk.org/the-european-convention-of-human-rights-and-climate-change-finally/>
'Permanent Peoples' Tribunal' <https://permanentpeoplestribunal.org/?lang=en>
Perry, Simona, 'Self-Determination and the Right to Information on the Shale Gas Frontier', *Center for Humans and Nature* (3 March 2017)
Peters, Birgit, 'The European Court of Human Rights and the Environment' in Edgardo Sobenes, Sarah Mead and Benjamin Samson (eds), *The Environment through the Lens of International Courts and Tribunals* (T.M.C. Asser Press, 2022) 189
Piesing, Mark, 'How to Build a Nuclear Warning for 10,000 Years' Time', *BBC Future* (online, 3 August 2020) <www.bbc.com/future/article/20200731-how-to-build-a-nuclear-warning-for-10000-years-time>
Preston, Christopher J, 'Re-thinking the Unthinkable: Environmental Ethics and the Presumptive Argument against Geoengineering' (2011) 20(4) *Environmental Values* 457
Quiggin, John, 'Peter Dutton Has Promised to Solve Our Energy Problems – but His Nuclear Policy Still Leaves Australians in the Dark', *The Conversation* (19 June 2024) <http://theconversation.com/peter-dutton-has-promised-to-solve-our-energy-problems-but-his-nuclear-policy-still-leaves-australians-in-the-dark-232816>
Rabone, Muriel et al, 'How Many Metazoan Species Live in the World's Largest Mineral Exploration Region?' (2023) 33(12) *Current Biology* 2383
Ramana, MV, 'Why Technical Solutions Are Insufficient' in Achim Brunnengräber and Maria Rosaria Di Nucci (eds), *Conflicts, Participation and Acceptability in Nuclear Waste Governance: An International Comparison Volume III* (Springer Fachmedien, 2019) 21
Rao, Sahana, 'Delaware River Commission Further Guards Basin from Fracking', *Natural Resources Defence Council* (19 December 2022) <www.nrdc.org/bio/sahana-rao/delaware-river-commission-further-guards-basin-fracking>
Rawls, John, *A Theory of Justice* (Harvard University Press, 1971)
Justice as Fairness: A Restatement (ed Erin Kelly, Harvard University Press, 2001)
Raz, Joseph, *The Morality of Freedom* (Clarendon, 1986)
Redgwell, Catherine, *Intergenerational Trusts and Environmental Protection* (Juris Publishing, Manchester University Press, 1999)
Reed, Graeme et al, 'Indigenous Guardians as an Emerging Approach to Indigenous Environmental Governance' (2021) 35(1) *Conservation Biology* 179

Ridley, Matt, *The Shale Gas Shock* (Global Warming Policy Foundation, 2011) <www.thegwpf.org/images/stories/gwpf-reports/Shale-Gas_4_May_11.pdf>

Ro, Christine, 'The Staggering Timescales of Nuclear Waste Disposal', *Forbes* <www.forbes.com/sites/christinero/2019/11/26/the-staggering-timescales-of-nuclear-waste-disposal/>

Rodriguez, Jose, Joonghyeok Heo and Kee Han Kim, 'The Impact of Hydraulic Fracturing on Groundwater Quality in the Permian Basin, West Texas, USA' (2020) 12(3) *Water* 796

Rodriguez-Rivera, Luis, 'Is the Human Right to Environment Recognised under International Law? It Depends on the Source' (2001) 12 *Colorado Journal of International Environmental Law and Policy* 1

Rogers, Nicole, 'Activists Are Using the Climate Emergency as a New Legal Defence to Justify Law-Breaking', *The Conversation* <http://theconversation.com/activists-are-using-the-climate-emergency-as-a-new-legal-defence-to-justify-law-breaking-122949>

Ross, Andrew and H Damon Matthews, 'Climate Engineering and the Risk of Rapid Climate Change' (2009) 4(4) *Environmental Research Letters* 045103

Rost Rublee, Maria, 'AUKUS: What to Do with Nuclear Waste?', *The Lowy Institute* (17 April 2023) <www.lowyinstitute.org/the-interpreter/aukus-what-do-nuclear-waste>

Royal Society (Great Britain), *Geoengineering the Climate: Science, Governance and Uncertainty* (The Royal Society, 2009)

Russo, Antonello et al, 'Incidence of Childhood Leukemia before and after Shut Down of Nuclear Power Plants in Germany in 2011: A Population-Based Register Study during 2004 to 2019' (2023) 152(5) *International Journal of Cancer* 913

Schmid, Evelyne, 'Victim Status before the ECtHR in Cases of Alleged Omissions: The Swiss Climate Case', *EJIL: Talk!* (30 April 2022) <www.ejiltalk.org/victim-status-before-the-ecthr-in-cases-of-alleged-omissions-the-swiss-climate-case/>

Scientific Inquiry into Hydraulic Fracturing in the Northern Territory, *Final Report* (Northern Territory Government, 2018)

'Scientists Test Fukushima Fish after Nuclear Plant Water Release', *Reuters* (online, 19 October 2023) <www.reuters.com/world/asia-pacific/scientists-test-fukushima-fish-after-nuclear-plant-water-release-2023-10-19/>

Setzer, Joana and Catherine Higham, *Global Trends in Climate Change Litigation: 2024 Snapshot* (Grantham Research Institute on Climate Change and the Environment, London School of Economics and Political Science, 2024)

Shapovalova, Daria, 'Indigenous Rights and Resource Extraction: One Step Forward Two Steps Back', *Environmental Policy and Law* (7 December 2020) <https://environmentalpolicyandlaw.com/news-blog/indigenous-rights-and-resource-extraction-one-step-forward-two-steps-back>

Short, Damien et al, 'Extreme Energy, "Fracking" and Human Rights: A New Field for Human Rights Impact Assessments?' (2015) 19(6) *The International Journal of Human Rights* 697

Shrader-Frechette, Kristin, *What Will Work: Fighting Climate Change with Renewable Energy, Not Nuclear Power* (Oxford University Press, 2011)

Shue, Henry, 'Changing Images of Climate Change: Human Rights and Future Generations: Choosing a Future' (2014) 5 *Journal of Human Rights and the Environment* 50

'Subsistence Emissions and Luxury Emissions' (1993) 15(1) *Law and Policy* 39

The Pivotal Generation: Why We Have a Moral Responsibility to Slow Climate Change Right Now (Princeton University Press, 2021)

Singh, Pradeep A, 'Deep Seabed Mining and Sustainable Development Goal 14' in Walter Leal Filho et al (eds), *Life below Water* (Springer International Publishing, 2022) 271

'The Invocation of the "Two-Year Rule" at the International Seabed Authority: Legal Consequences and Implications' (2022) 37(3) *The International Journal of Marine and Coastal Law* 375

Slezak, Michael and Marty Smiley, 'Once a Decade Lake Eyre Comes Alive. But Scientists Say the Gas Industry's Final Push Is Putting It All at Risk', *ABC News* (online, 3 October 2022) <www.abc.net.au/news/2022-10-04/lake-eyre-impact-of-gas-development/101486164>

Slobodian, Lydia, 'Defending the Future: Intergenerational Equity in Climate Litigation' (2020) 32 *Georgetown Environmental Law Review* 569

'Obligations to Transgenerational Groups: A Justification for Sustainable Environmental Policy' (2012) 24 *Georgetown International Environmental Law Review* 387

Sovacool, Benjamin K, 'Cornucopia or Curse? Reviewing the Costs and Benefits of Shale Gas Hydraulic Fracturing (Fracking)' (2014) 37 *Renewable and Sustainable Energy Reviews* 249

Sovacool, Benjamin K et al, 'The Decarbonisation Divide: Contextualizing Landscapes of Low-Carbon Exploitation and Toxicity in Africa' (2020) 60 *Global Environmental Change* 102028

'Who Are the Victims of Low-Carbon Transitions? Towards a Political Ecology of Climate Change Mitigation' (2021) 73 *Energy Research & Social Science* 101916

Sovacool, Benjamin K and Michael H Dworkin, 'Energy Justice: Conceptual Insights and Practical Applications' (2015) 142 *Applied Energy* 435

Spielmann, Dean, 'Allowing the Right Margin: The European Court of Human Rights and the National Margin of Appreciation Doctrine: Waiver or Subsidiarity of European Review?' (2012) 14 *Cambridge Yearbook of European Legal Studies* 381

Stanner, WEH, 'Aboriginal Territorial Organization: Estate, Range, Domain and Regime' (1965) 36(1) *Oceania* 1

Stanner, WEH, *The Dreaming & Other Essays* (1956)

Stern, Nicholas, *The Economics of Climate Change: The Stern Review* (Cambridge University Press, 2007)

Svoboda, Toby, 'Is Aerosol Geoengineering Ethically Preferable to Other Climate Change Strategies?' (2012) 17(2) *Ethics and the Environment* 111

Szabó, Marcel, 'National Institutions for the Protection of the Interests of Future Generations' (2015) 2(2) *E-Publica: Revista Electronica de Direito Publico* 6

The Scottish Government, *Independent Expert Scientific Panel – Report on Unconventional Oil and Gas* (The Scottish Government, 2014)

'The Strasbourg Principles of International Environmental Human Rights Law – 2022' (2022) 13 *Journal of Human Rights and the Environment* 195

This Is What We Die For: Human Rights Abuses in the Democratic Republic of the Congo Power the Global Trade in Cobalt (Amnesty International, 2016) <www.amnesty.org/en/wp-content/uploads/2021/05/AFR6231832016ENGLISH.pdf>

'Thousands Rally in Australia against Potential Submarine Base', *Reuters* (online, 6 March 2023) <www.reuters.com/world/asia-pacific/thousands-rally-australia-against-potential-submarine-base-2023-05-06/>

Tigre, Maria Antonia and Natalia Urzola, 'The 2017 Inter-American Court's Advisory Opinion: Changing the Paradigm for International Environmental Law in the Anthropocene' (2021) 12(1) *Journal of Human Rights and the Environment* 24

Tigre, Maria Antonia et al, 'A Request for an Advisory Opinion at the Inter-American Court of Human Rights: Initial Reactions', *Climate Law Blog* (17 February 2023) <https://blogs.law.columbia.edu/climatechange/2023/02/17/a-request-for-an-advisory-opinion-at-the-inter-american-court-of-human-rights-initial-reactions>

Tigre, Maria Antonia and Maxim Bönnemann, 'The Transformation of European Climate Change Litigation: Introduction to the Blog Symposium', *Climate Law Blog* (9 April 2024) <https://blogs.law.columbia.edu/climatechange/2024/04/09/the-transformation-of-european-climate-change-litigation-introduction-to-the-blog-symposium/>

Tokarz, Stephen M, 'A Golden Opportunity Dismissed: The New Zealand v. France Nuclear Test Case' (1998) 26(4) *Denver Journal of International Law and Policy* 745

Tremmel, Jörg (ed), *Handbook of Intergenerational Justice* (Edward Elgar, 2006)

Trickey, Kevin, Nicholas Hadjimichael and Prachi Sanghavi, 'Public Reporting of Hydraulic Fracturing Chemicals in the USA, 2011–18: A before and after Comparison of Reporting Formats' (2020) 4(5) *The Lancet Planetary Health* e178

Troy, Jakelin, Laura Rademaker and Ann McGrath, *Everywhen: Australia and the Language of Deep History* (NewSouth Publishing, 2023)

Tsubokura, Masaharu, 'Secondary Health Issues Associated with the Fukushima Daiichi Nuclear Accident, Based on the Experiences of Soma and Minamisoma Cities' (2018) 67(1) *Journal of the National Institute of Public Health* 71

United Nations, 'Summit of the Future: Multilateral Solutions for a Better Tomorrow' (United Nations, 2024) <www.un.org/en/summit-of-the-future>

United Nations Environment Programme, *Emissions Gap Report 2023: Broken Record – Temperatures Hit New Highs, yet World Fails to Cut Emissions (Again)* (20 November 2023) <www.unep.org/resources/emissions-gap-report-2023>

United Nations Environment Programme, *Rio Declaration on Environment and Development* (No A/Conf.151/26, 14 June 1992)

United Nations Scientific Committee on the Effects of Atomic Radiation, 'Assessments of the Radiation Effects from the Chornobyl Nuclear Reactor Accident' (2008) <www.unscear.org/unscear/en/areas-of-work/chernobyl.html>

United Nations Secretary-General, *Intergenerational Solidarity and the Needs of Future Generations* (General Assembly 68th sess, 5 August 2013) <https://sustainabledevelopment.un.org/content/documents/2006future.pdf>

United States Environmental Protection Agency, *Hydraulic Fracturing for Oil and Gas: Impacts from the Hydraulic Fracturing Water Cycle on Drinking Water Resources in the United States (Final Report)* (Reports & Assessments, December 2016)

'U.S. Activists Arrested at European Air Bases Protesting U.S. Nuclear Weapons Stationed There', *Democracy Now!* <www.democracynow.org/2023/8/10/nuclear_protests_netherlands>

van Dijk, Nicky, 'The Capability Approach as a Road Map for Re-Thinking Intergenerational Justice' in Jan Linehan and Peter Lawrence (eds), *Giving Future Generations a Voice: Normative Frameworks, Institutions and Practice* (Edward Elgar, 2021) 42

Van Dover, Cindy Lee, 'Mining Seafloor Massive Sulphides and Biodiversity: What Is at Risk?' (2011) 68(2) *ICES Journal of Marine Science* 341

Van Dover, CL et al, 'Scientific Rationale and International Obligations for Protection of Active Hydrothermal Vent Ecosystems from Deep-Sea Mining' (2018) 90 *Marine Policy* 20

Van Dyke, Vernon, 'Collective Entities and Moral Rights: Problems in Liberal-Democratic Thought' (1982) 44(1) *The Journal of Politics* 21

Vanreusel, Ann et al, 'Threatened by Mining, Polymetallic Nodules Are Required to Preserve Abyssal Epifauna' (2016) 6(1) *Scientific Reports* 26808

Vidal, John and environment editor, 'Not Guilty: The Greenpeace Activists Who Used Climate Change as a Legal Defence', *The Guardian* (online, 10 September 2008) <www.theguardian.com/environment/2008/sep/11/activists.kingsnorthclimatecamp>

Watts, Jonathan, 'How the Race for Cobalt Risks Turning It from Miracle Metal to Deadly Chemical', *The Guardian* (online, 18 December 2019) <www.theguardian.com/global-development/2019/dec/18/how-the-race-for-cobalt-risks-turning-it-from-miracle-metal-to-deadly-chemical>

Welsh Government, 'Licensing Powers on "Fracking" Transferred to Wales' (online, 1 October 2018) <https://gov.wales/licensing-powers-fracking-transferred-wales>

'What Is Fracking and Why Is It Controversial?', *BBC News* (online, 26 October 2022) <www.bbc.com/news/uk-14432401>

Whyte, Kyle, 'Settler Colonialism, Ecology, and Environmental Injustice' (2018) 9(1) *Environment and Society* 125

Whyte, Kyle Powys, 'Food Sovereignty, Justice, and Indigenous Peoples' in Anne Barnhill, Mark Budolfson and Tyler Doggett (eds), *The Oxford Handbook of Food Ethics* (Oxford University Press, 2018) 345

Witter, Roxana et al, *Potential Exposure-Related Human Health Effects of Oil and Gas Development: A White Paper* (Colorado School of Public Health, 2008)

Wolfe, Karrie, 'Greening the International Human Rights Sphere? Environmental Rights and the Draft Declaration of Principles on Human Rights and the Environment' (2003) 9(1) *Appeal: Review of Current Law and Law Reform* 45

'Woman Arrested in SA Nuclear Protest', *SBS News* (6 May 2016) <www.sbs.com.au/news/article/woman-arrested-in-sa-nuclear-protest/kojdfglgf>

World Commission on Environment and Development, *Our Common Future: Report of the World Commission on Environment and Development* (No A/42/427, 1987)

World Health Organization, *1986–2016: Chernobyl at 30: An Update* (25 April 2016)

World Nuclear Association, *World Nuclear Performance Report 2022* (2022) <www.world-nuclear.org/getmedia/9dafaf70-20c2-4c3f-ab80-f5024883d9da/World-Nuclear-Performance-Report-2022.pdf.aspx>

World Nuclear Association, 'Emerging Nuclear Energy Countries' (2023) <https://world-nuclear.org/information-library/country-profiles/others/emerging-nuclear-energy-countries.aspx>

World Nuclear Association, 'Nuclear Power in the World Today' (August 2023) <https://world-nuclear.org/information-library/current-and-future-generation/nuclear-power-in-the-world-today.aspx>

World Nuclear Waste Report (2019) <https://worldnuclearwastereport.org/>

Yamaguchi, Mari, 'Japan Top Court: Government Not Responsible for Fukushima Disaster', *The Diplomat* (online, 17 June 2022) <https://thediplomat.com/2022/06/japan-top-court-government-not-responsible-for-fukushima-disaster/>

Young, Stephen, *Indigenous Peoples, Consent and Rights: Troubling Subjects* (Routledge, 2020)

LEGISLATION

Climate Change Amendment (Duty of Care and Intergenerational Climate Equity) Bill 2023(Australia)

Constitution of Ecuador 2008

Constitution of Kenya 2010

The Constitution of the State of Montana (Annotated) 2023

INTERNATIONAL INSTRUMENTS

Committee on Economic Social and Cultural Rights, *General Comment No. 12: The Right to Adequate Food (Art. 11)* (1999)

Committee on Economic Social and Cultural Rights, *General Comment No. 14: The Right to the Highest Attainable Standard of Health* (2000)

Committee on Economic Social and Cultural Rights, *General Comment No. 15: The Right to Water* (2003)

Committee on the Rights of the Child, *General Comment No. 26 on Children's Rights and the Environment, with a Special Focus on Climate Change* (2023)

Draft Articles on Diplomatic Protection 2006

Human Rights Committee, *General Comment No. 31: The Nature of the General Legal Obligation Imposed on States Parties to the Covenant* (29 March 2004)

Human Rights Committee, *General Comment No. 36: Article 6 (the Right to Life)* (30 October 2018)

Human Rights Committee, 'Rules of Procedure of the Human Rights Committee' <http://hrlibrary.umn.edu/hrcommittee/HRC-RULE.htm>

Human Rights Council, 'The Human Right to a Safe, Clean, Healthy and Sustainable Environment' <https://documents-dds-ny.un.org/doc/UNDOC/LTD/G21/270/15/PDF/G2127015.pdf?OpenElement>

International Law Commission, *Draft Articles on Responsibility of States for Internationally Wrongful Acts, with Commentaries* (2001)

International Law Commission, *Draft Articles on the Prevention of Transboundary Harm from Hazardous Activities* (2008)

International Seabed Authority, 'Draft Regulations on Exploitation of Mineral Resources in the Area: Consolidated Text' (2004) <www.isa.org.jm/wp-content/uploads/2024/02/Consolidated_text.pdf>

Kioa Climate Emergency Declaration (2022)
Maastricht Principles on the Human Rights of Future Generations (2023)
Office of the High Commissioner for Human Rights, *Guiding Principles on Business and Human Rights: Implementing the United Nations "Protect, Respect and Remedy" Framework* (2011) <www.ohchr.org/documents/publications/GuidingprinciplesBusinesshr_eN.pdf>
Responsibility of States for Internationally Wrongful Acts 2001
Rio+20 Conference on Sustainable Development, *The Future We Want (Outcome Document)* (2012) <https://sustainabledevelopment.un.org/rio20/futurewewant>
United Nations General Assembly, *Request for an Advisory Opinion from the International Court of Justice on the Legality of the Threat or Use of Nuclear Weapons*, GA Res 49/75K (9 January 1995)
United Nations General Assembly, *Universal Declaration of Human Rights*, GA Res 217A(III) (10 December 1948)
United Nations Conference on the Environment, Declaration of the United Nations Conference on the Human Environment (Stockholm Declaration) (1972)
United Nations General Assembly, *The Human Right to a Clean, Healthy and Sustainable Environment*, GA Res 76/300 (28 July 2022)
United Nations Human Rights Council, *The Human Right to a Clean, Healthy and Sustainable Environment*, Res 48/13 (18 October 2021)
United Nations Summit of the Future, *Pact for the Future, Global Digital Compact and Declaration for Future Generations* (Outcome Documents of the UN Summit of the Future, 22–23 September 2024)
United Nations Summit of the Future, *Zero Draft of the Declaration on Future Generations* (26 March 2024) <www.un.org/sites/un2.un.org/files/co-facilitators_zero_draft_of_the_declaration_on_future_generations_26_march_2024_final_.pdf>

Index